T0310821

PROBABILITY, INFORMATION, AND PHYSICS

Problems with Quantum Mechanics in the
Context of a Novel Probability Theory

World Scientific Series in Information Studies
(ISSN: 1793-7876)

Published:

World Scientific Series in Information Studies — **Vol. 16**

PROBABILITY, INFORMATION, AND PHYSICS

Problems with Quantum Mechanics in the Context of a Novel Probability Theory

Paolo Rocchi

IBM, Italy & LUISS University, Italy

World Scientific

NEW JERSEY · LONDON · SINGAPORE · BEIJING · SHANGHAI · HONG KONG · TAIPEI · CHENNAI · TOKYO

Published by

World Scientific Publishing Co. Pte. Ltd.

5 Toh Tuck Link, Singapore 596224

USA office: 27 Warren Street, Suite 401-402, Hackensack, NJ 07601

UK office: 5 / Shelton Street, Covent Garden, London WC2H 9HE

Library of Congress Control Number: 2023022321

British Library Cataloguing-in-Publication Data
A catalogue record for this book is available from the British Library.

World Scientific Series in Information Studies — Vol. 16
PROBABILITY, INFORMATION, AND PHYSICS
Problems with Quantum Mechanics in the Context of a Novel Probability Theory

ISBN 978-981-127-274-5 (hardcover)
ISBN 978-981-127-275-2 (ebook for institutions)
ISBN 978-981-127-276-9 (ebook for individuals)

For any available supplementary material, please visit
https://www.worldscientific.com/worldscibooks/10.1142/13319#t=suppl

Desk Editors: Balasubramanian Shanmugam/Tan Rok Ting

Typeset by Stallion Press
Email: enquiries@stallionpress.com

Printed in Singapore

"Simplicity is the ultimate sophistication."
Leonardo da Vinci

"The essential content (...) of course, does not lie in the equations;
it lies in the ideas that lead to those equations"
Edwin Thompson Jaynes

About the Author

Upon retirement Paolo Rocchi has been recognized as an Emeritus Docent at IBM for his scientific achievements and nowadays serves as an Adjunct Professor at Luiss University of Rome. He has made contributions to a wide range of research areas including information theory, reliability theory, software engineering, quantum mechanics, and computer science education. He authored over 140 works including a dozen volumes. Rocchi has the reputation of being a solitary scout in search of new pathways and is not a stranger to groundbreaking proposals: this book is in line with his style.

About the Book

This book includes three main sections. The first and second together present a theory of probability that aims to unify the leading interpretations presented in the literature: frequentist, Bayesian, logical, etc. The third section uses the comprehensive theory to answer complex questions in quantum mechanics that have long been debated. The entire book proposes original and intriguing solutions that several experimental cases substantiate.

Contents

Abbreviations and Acronyms

BN	Bayesian Networks
CD	Criteria for Discernment
CF	Computational Formula
CM	Classical Mechanics
CT	Classical Theorem
DF	Definitional Formula
LLN	Law of Large Numbers
P	Probability
PU	Principle of Uniqueness
QM	Quantum Mechanics
SL	Structure of Levels
TA	Theorem of Addition
TC	Theorem of Continuity
TCS	Theorem of the Complete Structure
TD	Theorem of Discontinuity
TI	Theorem of Irreversibility
TIC	Theorem of Initial Conditions
TIS	Theorem of the Incomplete Structure
TLC	Theorem of Logical Compatibility
TLN	Theorem of Large Numbers
TM	Theorem of Multiplication
TMC	Theorem of Minimal Configuration

TR	Theorem of Reduction
TSN	Theorem of a Single Number
TUBL	Theorem of Upper-Bound Limit
VC	Validity Criterion

Introduction

The last decades have witnessed the true union of technology and human life for the first time on a mass scale. People in every part of the world are using technology to travel, communicate, learn, do business, live comfortably, and do other pleasant things. Our lifestyle relies on a variety of devices that ensure prosperity and welfare.

In this landscape, mathematicians play a role which is crucial and silent at the same time as they work in the 'second line'. They do not provide technology-based products, but the calculus tools needed by the engineers and professionals who manufacture these products. In broad strokes, customers depend on the doers of the technological society and in turn, the latter rely on mathematicians. If mathematics fails, common people can perceive the fiasco, since the effects of mathematicians' actions multiply in positive terms and negative equally. Victories are great, and defeats can be too.

In modern times, probability and statistics support executives, politicians, managers, decision-makers, and many other influential persons. Probability and statistics have become familiar even to laymen who are aware of economic trends, consumers' choices, the consensus to a political determination, etc. The patient is made conscious of the risk related to his or her illness; the investor forecasts his earnings chance, and exit polls predict voters' decisions.

Despite the popularity of statistical reports, the meanings of numbers do not sound perfectly clear. For some, probability is a wish; for another, it is solid data; for others, a possibility; for someone else, a guess, a supposition, etc. Confusion reigns supreme over this subject, and perhaps quantum scientists suffer the most critical consequences of the defeat of mathematicians. The behaviors of particles are fundamentally probabilistic and still appear to be so unintelligible and inexplicable.

This work is an attempt to explore the fundamentals of probability calculus and to use them in modern physics. The following pages will look into the ideas, methods, and criteria shared by probabilists. The second part puts forward a theoretical framework that aims to unify and integrate the leading interpretations of probability which we read in the literature. The third part of the book proposes new answers to go beyond the present stagnant state of quantum mechanics. The three parts are closely linked: the first is necessary to access the second part and this, to interpret the last topics.

Part 1

What Went Wrong in Probability Theory?

Chapter 1

Pascal's Manifesto

Frequently, gamblers are superstitious, and in the past they were familiar with talismanic gestures and signs to avoid bad luck. Their cultural background was often characterized by belief in supernatural influence, extracorporeal perceptions, and strange ideas about fate (Hald, 2003).

Over the centuries, gamblers made attempts to improve their chances of winning. They endeavored to obtain more serious forecasts even if historians have not found evidence of conscientious mathematical studies (Hacking, 2006). Let us cast a glance at two books and Pascal's letters which provide the earliest bibliographical testimonies of calculations in gambling.

1.1 Two Books

De Vetula (On the Old Woman) recounts an imaginary autobiography of the Roman poet Ovid. Researchers still ignore the authentic author of this booklet, written in about the 13th century. *De Vetula* could be classified as a morality poem, which means converting the reader to an honest lifestyle.

The first chapter describes the juvenile life of Ovid delighted by country pursuits: hunting, dice games, fishing, chess, etc. A long passage dealing with dice-playing illustrates a table that features the 216 possible outcomes of three dice. The author offers the oldest systematic illustration of a chance game. It may be said that he carried out the earliest documented combinatory analysis (Bellhouse, 2000).

The anonymous author had literary and moralistic interests, he exhibits imagination and creativity, and supports the shared idea that the calculus of probability flourished among people far removed from rigorous mathematical preparation.

Jerome Cardan (1500–1571) was an Italian chronic gambler, philosopher, astrologer, inventor, and even mathematician. Around the years 1560–1564, he wrote the *Liber de Ludo Aleae* (The Book on Games of Chance), which was published posthumously in 1663 for reasons that have not been clarified so far. Cardan proves to be cognizant of the following mathematical statements:

(i) He demonstrates the efficacy of defining odds as the ratio of favorable outcomes to total outcomes and shows to be familiar with the equation that nowadays is often called the *classical formula* of probability calculus.

(ii) He maintains that if the probability of the first player winning the wager is p, then the probability of winning for the second is $(1-p)$; somehow Cardan introduces the *addition rule*.

(iii) If p is the chance that an event happens in one experiment, the chance of it happening twice is p^2; for three successes the formula should be p^3 and so forth. He makes us understand that he is cognizant of the *multiplication rule*.

(iv) Cardan holds that the accuracy of predictions tends to improve with the number of trials. That is, he was aware of the fundamental rule currently referred to as the *law of large numbers*.

Liber offers the most accurate account of the calculus of probability written until then (Bellhouse, 2005). The author devotes several pages to games with dice and cards, including a miscellany of mathematics, philosophy, methods of cheating, tips, and religious remarks (Waters, 2007). The headings of some paragraphs give an idea of the variety of subjects treated: "Why gambling was condemned by Aristotle", "Who should play and when", and "Do those who teach also play well?". This assortment of topics presents evidence that the expressions (i)–(iv) were employed in contexts which were rather apart from pure mathematics.

1.2 Pascal's Letters

Following the history of probability, we find a book by Galileo Galilei (1564–1642) and other documents, but we confine ourselves to the correspondence of Blaise Pascal (1623–1662), including seven letters and a report (Ore, 1960).

In the year 1654 the nobleman Antoine Gombaud posed a pair of questions to Pascal, which may be summed up this way:

a. Is it more probable to get a six throwing a die four times or to get double six rolling two dice twenty-four times?
b. If two players suddenly interrupt a game and have won different points, how should the stake be fairly divided?

The personal curiosity of Gombaud raised question **a**, and Pascal proved that the first outcome is slightly more probable than a double six in 24 throws.

Problem **b** was of significantly greater importance and involves gamblers even nowadays. The interruptions of a game can arise from the decision of a player who is losing money, while the winning one wants to continue. The diverging interests can trigger quarrels and disputes that easily degenerate. Luca Pacioli (about 1445–1517) suggested subdividing the total amount of the stake into parts proportional to the scores of the players at the time of the interruption. However, Pacioli overlooked the possible evolution of the play that harms one player and benefits the other (Edwards, 1983). This detail makes us aware of the sophisticated problems tackled by amateur mathematicians, especially wealthy gamblers, who searched for the complete answer in the 17th century.

Pascal employed a recursive method but remained doubtful about his solution and sought confirmation from Pierre de Fermat (1601–1665), who confronted issue **b** following a different route. The two Frenchmen employed dissimilar criteria, yet they reached the same conclusion. The advanced and consistent approaches, which they adopted, demonstrated that the calculus of probability had a solid logical base, and enlightened Pascal.

In the autumn of that same year 1654, following an invitation of the Académie Parisienne, Pascal prepared a report illustrating his current scientific investigations. The letter closes with this passage:

> "(. . .) the knowledge of random events roamed hesitant so far, instead nowadays those facts which rebel to the experiment, do not avoid the rationality of the rules. And we have treated those random phenomena by means of exact calculations which share the mathematical certainty and are boldly progressing. By joining the rigor of scientific demonstrations with the uncertainty of chances and by reconciling these two apparently contradictory items, we can assign the following amazing title: *Geometry of Chance* to the discipline embracing both of them." (Chevalier, 1950) (Translation and Italics are mine)

The letter presented a striking list of the research-topics addressed by Pascal: numerical analysis, conics, the geometry of solids, the theory of perspective, studies about absolute and partial vacuum, and others. He devoted only a few lines or a few words to each of these numerous subjects while the probability calculus occupied almost one-third of the letter. The ample presentation reveals the fervor of the scientist for the discovery of the new field of mathematics which had been the 'property' of bettors and gamblers until then. Pascal claimed with unconcealed enthusiasm that the concept of chance could give substance to an original area of study, and from then onward, probability became the key to treating uncertainty.

1.3 What About the Geometry of Chance?

Pascal issued his *manifesto* with lively interest but did not devote much energy to it. Shortly after, he had a mystical experience and gave up all scientific inquiry. Philosophical and religious arguments absorbed his mind. Eight years after the mentioned report, he died, at the age of 39.

Despite Pascal's defection, mathematicians from various countries tackled application problems of increasing complexity. Historians illustrate the popularity of those demanding issues, and also a surprising divergence which came to light after the first decades. Probability was no longer the obscure territory of gamblers and amateur mathematicians, it had become an official mathematical sector.

Applied studies achieved brilliant results; nevertheless, theoretical inquiries began somewhat late, they advanced with difficulty, and are still in trouble.

Let us recall the earliest milestones of this mathematical sector.

Christian Huygens wrote 'De Ratiociniis in Ludo Aleae' in 1657, just 3 years after Pascal's report, which was the first systematic treatise on probability mostly focusing on gamblers' problems. Jacob Bernoulli devised the theorem currently labeled as the law of large numbers in *Ars Conjectandi* published posthumously in 1713. His nephew Daniel Bernoulli authored *Specimen Theoriae Novae de Mensura Sortis* (1738), where he described the St. Petersburg paradox. Abraham de Moivre introduced methods of approximating the binomial distribution in the book *The Doctrine of Chances* (1718) and applied his findings to gambling problems and actuarial tables. In the year 1733, Georges-Louis Leclerc, Comte de Buffon, solved the earliest problems in geometric probability. Thomas Bayes showed the relation between conditional probability and its reverse form. His paper 'An Essay towards solving a Problem in the Doctrine of Chances', presenting his famous theorem, was published two years after his death, in 1763.

This concise chronicle is enough to manifest how scholars went on in no particular order. They took the base concepts as intuitive ideas for a long while, nobody dared to formulate precise definitions for over 150 years. Pierre-Simon Laplace was the first to confront the question which was impossible to avoid:

What is probability?

In the second volume of the monumental treatise *Théorie Analytique des Probabilités* (1812), Laplace proposes an organic construction that fixes probability as the ratio of the number of favorable cases to the number of all possible cases. This definition presumes that the events under consideration are equiprobable; thereby it presents a logical vicious of circularity so evident that the French author remedied it by invoking the *principle of insufficient reason*, a correction which was not taken as fully satisfactory. A precise account of this treatise, whose subsequent editions were progressively enriched with new topics, may be found in (Molina, 1930).

In the 19th and 20th centuries, the scientific community increasingly felt the need for a solid foundation for the theorization of probability. The reader can get an idea of the impressive endeavors made by the authors from the classical study (Todhunter, 1865) up to most recent contributions which provide general views (Hald, 2003; Stigler, 1990; Bernstein, 1996; Vallverdú, 2011) or review individual schools of thought (Daston, 1994; Lehmann, 2011; Mc Grayne, 2012) or look into some special topics (Plackett, 1972; Gorroochurn, 2016). Researchers are still very prolific, but despite the huge ensemble of works, the profile of the probability theory remains substantially unclear. Quite a number of alternative constructions have been devised but none has reached universal consensus. The 'Geometry of Chance' dreamed by Pascal is not yet born.

This field of mathematics has proved to be incredibly powerful, practical, and successful in numerous areas. There seems to be no limit to its versatility in a considerable set of topics, yet the foundational arguments remain among the most controversial in mathematics. Over three centuries after Pascal's enthusiastic plan, we face a rich and popular field of mathematics which however has a reputation for being confusing, if not outright impenetrable from the conceptual viewpoint. Several eminent authors have verbalized their skeptical vision:

> "Probability is one of the outstanding examples of the 'epistemological paradox' that we can successfully use our basic concepts without actually understanding them" (Weizsäcker, 1973).

In a lecture, Bertrand Russel said:

> "Probability is the most important concept in modern science, especially as nobody has the slightest notion what it means" (cited by (Bell, 1945)).

Further difficulties arise when the calculus tackles a variety of different systems, which burdens the already confused basic notions with distinct and often conflicting philosophical connotations. Two experts may obtain the same numerical value but assign meanings to it that are impossible to reconcile. A comprehensive theory is delaying, and it is natural to wonder:

Why have mathematicians failed so far?

1.4 The Thesis of this Book

Science and technology are moving forward at a fast pace and underpin much of the progress of human welfare and civilization. The great benefits of these titanic efforts lie before our eyes. The work conducted by scientists and scholars is enormous. Nowadays, there are more scientists, more funding for science, and more scientific articles published than ever before (Krohn *et al.*, 1978). The defeat of probabilists and statisticians manifests a striking contrast in the modern cultural context and sounds somewhat incomprehensible:

What went wrong?

It is my thesis that Pascal and his followers had a false impression from the brilliant solutions to applied problems. Reading the 'manifesto' of the French mathematician, it is apparent how he imagined employing self-evident tenets similar to the elements of classical geometry. Researchers inherited this idea; they felt that the formalization of probability is simple or can be simplified. They kept this conviction even if, with the passage of time, they became conscious of the heavy web of notions underpinning this domain of knowledge. I do not mean to say the authors sustained fallacies, neither did they make wrong conclusions or assert false statements, rather I am inclined to hold that the experts who had the purpose of erecting the 'Geometry of Chance' regarded it as a somewhat simple construct and adopted inappropriate criteria of research.

The following pages throw out considerations in support of this thesis. The introductory Chapter 2 recalls the numerous and prismatic base tenets and some arguments debated by experts. Chapter 3 analyzes the minimalist perspective shared by probability theorists, and Chapter 4 analyzes some key elements that the literature neglects.

References

Bell E.T. (1945). *The Development of Mathematics* (McGraw Hill, New York).

Bellhouse D.R. (2000). De Vetula: A medieval manuscript containing probability calculations, *International Statistical Review*, 68(2), 123–136.

Bellhouse D.R. (2005). Decoding Cardano's Liber de Ludo Aleae, *Historia Mathematica*, 32(2), 180–202.

Bernstein P.L. (1996). *Against the Gods: The Remarkable Story of Risk* (Wiley & Sons, New York).

Chevalier J. (ed.) (1950). *Oeuvres Complétes de Blaise Pascal* (Gallimard, Paris).

Daston L. (1994). How probabilities came to be objective and subjective, *Historia Mathematica*, 21, 330–344.

Edwards A.W.F. (1983). Pascal's problem: The 'Gambler's Ruin', *International Statistical Review*, 51(1), 73–79.

Gorroochurn P. (2016). *Classic Topics on the History of Modern Mathematical Statistics: From Laplace to More Recent Times* (Wiley & Sons, New York).

Hacking I. (2006). *The Emergence of Probability: A Philosophical Study of Early Ideas about Probability, Induction and Statistical Inference*, 2nd edition (Cambridge University Press, Cambridge).

Hald A. (2003). *A History of Probability and Statistics and their Applications before 1750* (Wiley Interscience, Hoboken, NJ).

Krohn W., Layton Jr. E.T., and Weingart P. (eds.) (1978). *The Dynamics of Science and Technology: Social Values, Technical Norms and Scientific Criteria in the Development of Knowledge* (Reidel Publishing Company, Dordrecht).

Lehmann E.L.L. (2011). *Fisher, Neyman, and the Creation of Classical Statistics* (Springer, Berlin; New York).

Mc Grayne S.B. (2012). *The Theory That Would Not Die* (Yale University Press, New Haven, CT).

Molina E.C. (1930). The theory of probability: Some comments on Laplace's Théorie Analytique, *Bulletin American Mathematical Society*, 36(6), 369–392.

Ore O. (1960). Pascal and the invention of probability theory, *American Mathematics Monthly*, 67, 409–419.

Plackett R.L. (1972). Studies in the history of probability and statistics. XXIX: The discovery of the method of least squares, *Biometrika*, 59(2), 239–251.

Stigler S.M. (1990). *The History of Statistics: The Measurement of Uncertainty before 1900* (Belknap Press/Harvard University Press, Cambridge, MA).

Todhunter I. (1865). *A History of the Mathematical Theory of Probability from the Time of Pascal to that of Laplace* (MacMillan, London; New York).

Vallverdú J. (2011). History of probability, In M. Lovric (ed.), *International Encyclopedia of Statistical Science* (Springer, Berlin; New York).

von Weizsäcker C.F. (1973). Probability and quantum mechanics, *British Journal for the Philosophy of Science*, 24(4), 321–337.

Waters W.G. (2007). *Jerome Cardan: A Biographical Study* (Kessinger Pub. Co., Whitefish, MT).

Chapter 2

Easy to Use, Namely Substantially Knotty

The term *'probability'* (*P*) took its factual origin in judicial contexts as proved by the very etymology of the word *'probable'* deriving from the Latin *'probabilis'* that stands for something *provable* or *believable* before a court (Harper, 2000). Still today lawyers and magistrates accurately examine the *probable* elements in a lawsuit. The more a cue or evidence is probable, the better it can be accepted, and the better it serves as proof of an accusation or a defense.

2.1 Self-evident Formulas

Men of law were strongly involved in the meticulous evaluations of witnesses and practical clues but did not succeed in being so precise as to employ numbers to qualify the elements useful to a judgment. Highly cultured and learned professionals have failed, whereas ignorant people have hit the target. Gamblers, bookmakers, punters, and other individuals *not so learned in mathematics discovered the basics of probability calculus.*

From the Renaissance onward, those who wagered with money became familiar with the expressions presented by Cardan, which we reread using modern formalism.

2.1.1 Players anticipated the *classical formula* of Laplace relating the favorable cases to the total number N of equally likely outcomes

$$P = \frac{Q}{N} \tag{2.1}$$

Gamblers summed up the probabilities of the results A and B that are incompatible

$$P(\text{A or B}) = P(\text{A}) + P(\text{B}). \tag{2.2}$$

They even grasped that the probability of occurrence of both the independent outcomes A and B is equal to the product of probabilities

$$P(\text{A and B}) = P(\text{A}) \bullet P(\text{B}). \tag{2.3}$$

Lastly, gamblers became convinced that the relative frequency is the counterpart of probability in the world. The higher the number of gambling rounds, the clearer this connection becomes in physical reality

$$\text{Probability} \blacktriangleleft ----\blacktriangleright \text{Frequency.} \tag{2.4}$$

2.1.2 Early players contrived (2.1)–(2.4) by intuition and not after mathematical preparation, of which we do not have any bibliographical evidence. Later, probabilists and statisticians deemed these formulas so immediate and precise as to ratify them without discussion. All this demonstrates how ancient and modern experts share the impression of an easy way. Laplace (1814) summarizes this common perception:

> "The theory of probability is basically nothing but good sense reduced to calculation; it allows us to assess with a precision that exact minds feel by a sort of instinct, without often being able to give a reason for it."

Expressions (2.1)–(2.4) sound self-evident, but the same cannot be said for the concepts underpinning those formulas and the following pages will recall some controversial topics. This book does not have within its scope the development of philosophical arguments. I suppose the reader has some familiarity with probability and statistics, and shall express what needs to be said in a concise manner.

2.2 Probability from the Theoretical Viewpoint

Let us take a look at the prismatic nature of the concept of probability which scholars have discovered with the passage of time.

2.2.1 The masters traditionally recognized for leading explanations present the following spectrum of interpretations (Rocchi, 2014, Appendix A).

a. *Algebraic-Logic theory* – George Boole (2009) *associates probability with the logic of human thought* and calculates P through a set of algebraic operations acting on propositions that may be true or false.

b. *Inferential theory* – For John Maynard Keynes (1921), probability qualifies the logical support for the hypothesis which yields a conclusion. He identifies degrees of P with *degrees of rational belief.*

c. *Frequentist theory* – The earlier proponents of this idea were Jacob Bernoulli. Richard von Mises (1928), together with Antoine Cournot, John Venn (2018), Joseph Bertrand (1889), and many others, defines the probability of an attribute as the *limiting value of the relative frequency* with which this attribute recurs in the indefinitely prolonged sequence of trials.

d. *Subjective and Bayesian theory* – Frank P. Ramsey (1931) and Bruno de Finetti (1931), together with Leonard J. Savage (1954) and other Bayesians, regard probability as a *personal degree of credence* in the occurrence of the random event.

e. *Inductive logical theory* – Harold Jeffrey (1939) sees P as a generalization of inductive logic, providing support for given data. Rudolf Carnap (1950) views probability theory as an extension of first-order logic, specifically as a logic of partial implication, and defines probability as the degree of confirmation of logical induction.

f. *Propensity theory* – Karl Popper (1959) interprets P as a *property of experiments which is objective and mind-independent.*

Another group of researchers conceives probability as an abstract parameter apart from any specific context:

1. *Classical theory* – Pierre Simon Laplace (1812) puts forward the *classical formula* (2.1) as the formal definition of P that presumes the cases are equally likely.

2. *Axiomatic theory* – Andrej Kolmogorov (1933) defines a measure space where P is a set-function of an abstract subset satisfying the axioms of *non-negativity, normalization,* and *finite additivity*.

3. *Quantum probability theory* – In the 1980s quantum mechanics led mathematicians to develop *a noncommutative theory* in order to clarify the special phenomena occurring in the quantum context (Meyer, 1993).

4. *Imprecise probability theory* – Peter Walley (1991) observed the partial precision characterizing the assessment of P and proposes a theory that replaces a single value with an interval specification that is determined by *lower* and *upper probabilities*.

5. *Qualitative probability theory* – One event turns out to be more likely than another without necessarily stating the exact probability values. First Sergei N. Bernstein and later other mathematicians have made attempts to axiomatize a qualitative approach (Gärdenfors, 1975) that offers a pragmatic, intuitive, and practical counterpoint to conventional probability.

6. *Negative probability theory* – Special phenomena occurring in the economy and quantum physics consist of coupled and mutually compensating events. They are formalized by means of positive and negative values of probability (Khrennikov, 2007; Burgin, 2016).

7. Finally, researchers formalize the probability properties by means of the *fuzzy theory* (Bugajski, 1996; Beer, 2009) and the *categories theory* (Giry, 1982). These efforts are conducted at high level of abstraction in order to measure P in a general space (Heunen, *et al.*, 2017).

2.2.2 Besides the 'orthodox' or 'radical' probabilists who claim only one view is valid, a large group of thinkers has a more ecumenical tendency. The members of this circle, whom I call 'dualist', tend to acknowledge a double interpretation of P. Various leading authors share this viewpoint; Ramsey argues about the difference between the notions of probability in physics and in logic, and admits the use of the frequentist model alongside the subjective model. Carnap recognizes *probability*$_1$ as a degree of confirmation and *probability*$_2$ that approaches the relative frequency. Popper introduces *long run* and *single case propensities*. Kolmogorov draws a separation between the mathematical formalization and the interpretations of probability and shows to be open to accept diverse viewpoints.

2.2.3 More recently, the *Bayesians* have inaugurated an interesting vein of research that tends to minimize the person's contribution to statistical inference by introducing reference priors or other constraints (Weisberg, 2011). Edwin Jaynes (Jaynes, 2003) claims that our initial credence function should be an accurate (and objective) description of how much information we have in hand. He qualifies information using the entropy of Shannon and introduces the *principle of maximum entropy* by stating that the probability distribution which best represents the current state of knowledge is the one with the highest entropy. *Objective Bayesian analysis* exploits prior distributions that express 'neutral' knowledge and leads to the posterior distribution, which should express the probability derived from the data alone (Berger, 2006).

Other attempts have been made to bring the different perspectives closer together. David Lewis (Lewis, 1976) formulates the so-called *principal principle* (PP) to optimize the link between the subjective and objective models. He argues that our degrees of belief should conform to our scientific knowledge; PP can be summarized by saying that if you know that the possibility of X is P, then you should believe X to the degree P. Philip Dawid (2004) strives to harmonize personal probabilities with the world out there using philosophical arguments. Richard Cox (1961) defines P as the measure of credibility consistent with logic. He shows that probability rules can be derived by applying logic and Boolean calculus. Another circle of dualists accepts distinct perspectives based on emerging pragmatic needs. Tukey (1960) and Gillies (2000) note that we can assign frequency probability to general laws of science and subjective probability to the management of an accidental case.

Nowadays, the dualist position seems to be gaining ground (Rocchi *et al.*, 2010). The writers have devised original ideas but have not proposed entirely convincing solutions so far. Trivially one could object why P should have two meanings and not three or four or more. It can be said that dualist authors tend to reconcile the different stances, they adhere to the positions listed from a. to f. although in more moderate terms.

2.2.4 Like some subtle scientific concepts, probability began with a seemingly straightforward, commonsensical meaning, which has become trickier as thinkers have taken a closer look at it.

The different theoretical interpretations, taken one by one, prove to be reasonable and even conform to the 'good sense' envisioned by Laplace. However, they do not present a coherent landscape and lead to an apparent conclusion: probability is a multifaceted concept that is difficult to be brought into a simple form, it is far from being an 'instinctive' notion as some presume.

2.3 Probability from the Empirical Viewpoint

Ancient gamblers had no doubt that the empirical frequencies converge toward the numerical values calculated in the abstract.

2.3.1 Jacob Bernoulli first described this mathematical property with the theorem that presently we call *law of large numbers* (LLN). Eminent authors improved it, Chebyshev, Markov, Borel, Cantelli, and others provided significant contributions. The *weak version* of LLN states that if X_1, X_2, X_3, $X_4 \ldots X_n$ are independent and identically distributed (i.i.d.) random variables, then the sample average $\bar{X}$ converges in probability to the theoretical mean μ

$$\overline{X} \xrightarrow{P} \mu, \quad as \ n \to \infty \tag{2.5}$$

In words, for any specified nonzero margin ε, no matter how small, there will be a very high probability that the mean of the observations $\overline{X}$ will be close to the expected value. The *strong version* proves that when the number of identically distributed randomly generated variables increases, the sample mean will almost surely approach μ

$$\overline{X} \xrightarrow{a.s.} \mu, \quad as \ n \to \infty. \tag{2.6}$$

Other works have improved our understanding of LLN, such as the *central limit theorem* and the *law of the iterated logarithm*, which specify the speed of convergence, and the *Glivenko–Cantelli theorem* on the convergence of the empirical distribution function.

2.3.2 Generally speaking, a mathematician does not care whether his abstract work can be employed in the living environment. He can create a construction that has no application, all the same his work is valid, provided it is consistent. For instance, Riemann, Beltrami, and Poincaré worked out non-Euclidean geometries and did not much care about the application of the new formulas, which at first glance seemed to have no nexus with practical experience.

The strong concern of theorists for empirical themes suggests that:

(1) Probability is a parameter deeply involved with practical topics.
(2) Probability opposes difficulties to experimental control.

Noteworthy problems indeed, in fact, various criticisms have been raised against LLN. The questions do not deal with the formal-mathematical aspects of the law, rather with topics that echo points **(1)** and **(2)** and involve 'logical' and 'metaphysical' arguments. For instance, Gnedenko and Kolmogorov (1954) note that

> The epistemological value of probability theory is based on the fact that chance phenomena, considered collectively and on a grand scale, create non-random regularity.

2.3.3 Why and how can outcomes that are considered independent, irregular, and in isolation be subjected to a common rule?

The literature reads a variety of problematic positions. For example, Popper faces the contradictory behavior of repeated unpredictable events that eventually provide outputs that are certain and arrives at the idea that events share a physical disposition or tendency that he calls 'propensity'. Henri Poincaré annotates that a theory based on randomness to some extent and consequently on our ignorance should not settle precise conclusions. Siméon Denis Poisson comments on the opposed tendencies of random events (Poisson, 1837) and Augustus de Morgan criticizes the reliability of predictions

"There is no prophecy of particular event in the theory of probabilities of which it is the very essence that there should be more or less tendency to falsehood in every one of its assertions. No result is announced, except as having a certain chance in its favour, which implies also a certain chance against it." (de Morgan, 1838)

Besides the skeptical writers, a group directly attacks the LLN such as Joseph Bertrand (1889) who wonders

"How dare we speak of the laws of chance? Is not chance the antithesis of all law?"

Steinhaus (1992) calls the special version of LLN proved by Émile Borel as "le paradoxe de Borel". Von Mises (1928) claims that one cannot infer statistical conclusions from premises expressing uncertainty; for him LLN cannot be a kind of logic bridge leading from uncertain assumptions to statistical conclusions.

2.3.4 By contrast, Jacob Bernoulli (2011) stresses the intuitive qualities of LLN, and notes that

"Sometimes the stupidest man by some instinct of nature per se and by no previous instruction (this is truly amazing) knows for sure" the law of large numbers.

The literature shows that the LLN, which, at first glance, seems to be inspired by 'some instinct', hides complicated and unresolved arguments. The concept of probability proves to be enigmatic in abstract and even more complicated on the practical plane.

2.4 Correlated Notions

An assortment of concepts involves the sustained attention of thinkers. *Chance, likelihood, randomness, uncertainty, possibility, probability,* and other notions resemble neighboring countries. Each is separate from the other, yet they share long borders and partially overlap.

2.4.1 The term '*chance*' is ordinarily associated with something that happens without a discernible or observable origin. Venn (2018) sees "chance as opposed to physical causation" and adds "causation is not necessary for that part of the process which belongs to probability." These considerations make the problem even more complex. If the concept of chance means the absence of any explicit reason, then the *principle of causality*, a cornerstone of science, technology, and logic, is denied (Illari and Russo, 2014).

2.4.2 Earlier discussion of causality dates back to Aristotle, who in his *Metaphysics* presents four distinct types of causes: *efficient, material, formal,* and *final.* The efficient cause proves to be near the perspective of modern scientists who ordinarily search for operational determinants. In fact, scientists cast light on every hidden sign and scrutinize the smallest indication in order to discover the causal factor of what they are studying (Skyrms, 1980).

The history of mankind bears testimony to noteworthy enterprises organized to look for the causes of defeats, financial ruins, catastrophic damages, pandemics, and so forth. Presently, 'Root Cause Analysis' (RCA) guides professionals to search for the out-of-sight reasons of crucial events. RCA was first applied in engineering, and then it was exploited in industry, business, medicine, and other fields (Okes, 2019; Robitaille, 2004).

Every sector of human activity treats the principle of causality as the central criterion to operate and understand. The concept of 'chance' as an unexpected fact occurring without plausible origin would present a paradoxical perspective which breaks up the logic of science and disregards the praxis of human wisdom. Probability calculus would be the unique sector that waives the explanation of the causes of phenomena.

2.4.3 The idea that chance should not have any precise cause sounds like a contradiction to the scientific logic and human intelligence. That is why the calculus of probability has raised the fierce criticism of *positivist* thinkers such as Auguste Comte (1995) who claims that "to evaluate a probability is a scientific fraud and a moral dishonesty." In particular, he stresses the limits of the ability of statistics to explain social phenomena and to describe the variable side of living beings.

2.4.4 Some authors are inclined toward a kind of 'trade-off'. They assume that chance has causes but these causes cannot be deciphered in a systematic manner and are of a type that does not constrain the future to a single course. Antoine Augustin Cournot introduces the conceptual separation between *linear causality* and *nonlinear causality*. In his theory of chances (1843) and later in the essay (1851), he defines chance as the meeting of several independent causal series. He looks at the intricate chains of interactions deriving from the encounter of different mechanisms that eventually result in a fortuitous event. Henri Poincaré (1916–1954) shares this perspective and argues that the global effect of tiny physical elements adds up over time and eventually embodies an aleatory phenomenon. Other thinkers theorize about the action of factors that are known in the abstract but cannot be measured in practice.

2.4.5 The study of chance becomes even more complicated when one recognizes that the detection of causes is naturally linked to human knowledge, as was already noted by Laplace (1814):

> "(...) we attribute the phenomena that seem to us and succeed without any particular order to changeable and hidden causes, whose action is designated by *the word chance, a word that after all is only the expression of our ignorance. Probability relates partly to this ignorance and in other parts to our knowledge.*" (Italics mine)

A large circle of modern probabilists, especially the subjectivists and the Bayesians, relate chance to human 'uncertainty' and 'defective cognition' (Mann, 1994; Teigen and Keren, 2020; Nickerson, 2002). For them chance is an artifact of our unawareness (Diaconis and Skyrms, 2018), and they have analyzed this intellectual inability from various viewpoints: some emphasize people's efforts to detect regular trends (Griffiths *et al.*, 2018); some think about the observable but unrepeatable outcomes; some evaluate the role of statistics and formal tools (Stern, 2011), and others argue about the involuntary or even voluntary actions that generate our ignorance (Werndl, 2009; Müller, 2019).

2.4.6 A group of scholars is inclined to explain the concept of chance by referring to correlated notions such as *randomness*. Antony Eagle (2018) cites the following commonplace claim:

"Something is random if and only if it happens by chance."

However, when the concepts of chance and randomness lean together, they do not clarify each other but instead tend to increase our own difficulties in comprehension.

Example. Bob, walking on a solitary island, finds some ancient golden treasure. Bob discovers the treasure by chance, but his involuntary and unrepeatable discovery cannot be defined as a random event.

For frequentists, the adjective 'random' stands for something that *lacks any clear rule*, and Kolmogorov asks:

> "Does the randomness of the event A demonstrate the absence of any law connecting the complex of conditions S and the event A?" (Aleksandrov *et al.*, 1999)

This question sounds rhetorical whereas the answer is anything but rhetorical. In fact, the widely shared idea which associates aleatory events with missing rules is not exactly true. A sequence of similar random occurrences can create an overall phenomenon that can be easily studied and forecast.

Example. Robert Brown observed the erratic behavior of a single grain of pollen in a liquid. The longer he waited, the further the particles moved away from their starting point, in a characteristic movement of increasing amplitude which we call *Brownian motion* (Mörters and Peres, 2010). The mean squared displacement of the particle with respect to the initial position x_0 can be established by the diffusion coefficient D of the particle

$$\overline{(x(t) - x_0)^2} = 2D\,(t - t_0). \tag{2.7}$$

2.4.7 It seems natural to relate chance and randomness to probability, yet the latter does not furnish solid support since the probability of patternless outputs does not exclude occasional regularity.

Example. Any sequence of k outcomes in heads and tails has this probability

$$P(k) = 2^{-k}. \tag{2.8}$$

The following strings present regular patterns and each one has the same probability as any other sequence of k tosses

"HHHHHHHHHHHHHH ... HH"

"TTTTTTTTTTTTTTTT ... TT"

"THTHTHTHTHTHTHTH ... TH"

The concept of probability does not guarantee that a regular configuration is absent.

2.4.8 The succinct survey presented in these pages is intended to cast light on the intricate web of issues lying at the base of indeterminism. Pascal believed that the 'Geometry of Chance' was close by; instead, the notions of probability, chance, randomness, uncertainty, etc. present multiple facets and are still up for debate. Probabilities are real numbers between zero and one, although this is not the same as deciding what probability actually is. Probability calculus provides us with an analytic tool of enormous power and scope, but the claims of Laplace and Bernoulli, who talk about 'good sense' and 'instinctive' ideas, do not seem to lie on solid ground.

2.5 Research Strategies

Exact science can be viewed as two trees with intertwined roots: one being theoretical, and the other being applied. The first tree basically addresses problems closely linked to practical situations; the second can be regarded as the ability to build up intellectual frameworks having rigorous internal logic. Theorization and applied calculus embody different processes and methodologies which sometimes present contrasting features.

2.5.1 *Theoretical and applied formulations* – There are mathematics-based disciplines that are underpinned by rather easy theoretical statements, while the applications challenge the experts.

Other disciplines offer simple methods for the calculus but are backed by demanding theoretical tenets.

Example. Two essential mechanisms regulate the flight of the planets around the Sun. Considering negligible interplanetary interactions, the gravitational and the centrifugal forces determine the orbits which have the form of an ellipse where the Sun occupies a focus

$$\frac{x^2}{a^2} + \frac{y^2}{b^2} = 1. \tag{2.9}$$

Undergraduate students easily acquire this simple model; instead, its application in astronomic observations prompts non-trivial efforts (Xu and Xu, 2013). The position of a planet in the sky requires the calculation of nine parameters at least (Table 2.1).

The probability field presents difficulties in inverse proportions to those of astronomy. The equations necessary to solve application problems turn out to be somewhat easy to learn (Section 2.1), whereas the underpinning tenets are anything but linear and form an intricate intellectual mesh (Sections 2.2, 2.3 and 2.4).

Table 2.1. The parameters necessary to find a planet in the sky.

1	The semi-major axis a
2	The eccentricity of the orbit
3	The inclination of the orbit
4	The position of the planet in its orbit at that date
5	The ecliptic longitude
6	The obliquity of the ecliptic
7	The distance between the perihelion and the ascending node of the orbit
8	The geographical longitude and latitude of the observer
9	The time zone of the observation

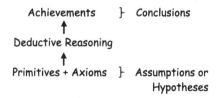

Figure 2.1. Conceptual map of a hypothetic-deductive theory.

2.5.2 *Direct theorization* – Theories are not discovered; they are created by authors who follow rational criteria. In particular, they assume straightforward initial tenets so that they can get any theoretical achievement by relating it to the hypotheses in a more or less direct way (Figure 2.1). The formal construct is based on *primitives* or *primary notions*, which Tarski calls 'undefined terms' because they are self-evident. The preliminaries also conform to Ernst Mach's *principle of economy* (a version of Occam's razor), which he summarizes this way: "Scientists must use the simplest means to arrive at their results" (Banks, 2004). Thus, a mathematician seeks the simplest starting point even when he intends to theorize a complicated concept (Dilworth, 2013).

2.5.3 Probability turns out to be far from being a univocal concept and is completely unsuitable for the role of a primitive element. This means that an author who intends to create a complete theory *cannot directly theorize P.*

When a theorist recognizes that the entity Y has many and varied characteristics, he does not begin 'the theory of Y' with Y, but with a simpler element. He does not directly formulate Y but commences with a more agile notion such as the argument X of $Y(X)$ or with the component Z of Y_Z, or he resorts to another intuitive subject that conforms with the role of a primitive.

The reader can note how the vast majority of the theorists, who have been mentioned in this book (and even others), *have made attempts to formulate P per se.* They directly attacked P without the aid of a plainer element, and this approach reveals how they underestimated the prismatic nature of probability. They did not openly confess that probability is simple, or can be simplified, but their shared research strategy unveils this concealed idea and provides evidence in favor of the thesis of this book.

References

Aleksandrov A.D., Kolmogorov A.N., and Lavrent'ev M.A. (1999). *Mathematics: Its Content, Methods and Meaning* (Dover Publications, Chicago).

Banks E. (2004). The philosophical roots of Ernst Mach's economy of thought, *Synthese*, 139(1), 23–53.

Beer M. (2009). Fuzzy probability theory, In R. Meyers (ed.), *Encyclopedia of Complexity and Systems Science* (Springer, Berlin; New York).

Berger J. (2006). The case for objective Bayesian analysis, *Bayesian Analysis*, 1(3), 385–402.

Bernoulli J. (2011). *Ars Conjectandi: Opus Posthumum* (Nabu Press, Charleston, SC).

Bertrand J. (1889). *Calcul des Probabilités* (Gauthier-Villard and Fils, Paris).

Boole G. (2009). *An Investigation of the Laws of Thought on Which are Founded the Mathematical Theories of Logic and Probabilities* (MacMillan, London; New York).

Bugajski S. (1996). Fundamentals of fuzzy probability theory, *International Journal of Theoretical Physics*, 35, 2229–2244.

Burgin M. (2016). Axiomatizing negative probability, *Journal of Advanced Research in Applied Mathematics and Statistics*, 1(1).

Carnap R. (1950). *Logical Foundations of Probability* (University of Chicago Press, Chicago).

Comte A. (1995). *Cours de Philosophie Positive: Quarantième Leçon* (Hermann, Paris).

Cournot A.A. (1843). *Exposition de la Théorie des Chances et des Probabilités* (Hachette, Paris).

Cournot A.A. (1851). *Essai sur les Fondements de nos Connaissances et sur les Caractères de la Critique Philosophique* (Hachette, Paris).

Cox R.T. (1961). *The Algebra of Probable Inference* (Johns Hopkins University Press, Baltimore, MD).

Dawid A.P. (2004). Probability, causality and the empirical world: A Bayes-de Finetti-Popper-Borel synthesis, *Statistical Science*, 19, 44–57.

de Finetti B. (1931). Sul Significato Soggettivo della Probabilità, *Fondamenta Matematicae*, 17, 298–329.

de Morgan A. (1838). *An Essay on Probabilities, and Their Application to Life Contingencies and Insurance Offices* (Longman, Orme, Brown, Green & Longmans, London).

Diaconis P. and Skyrms B. (2018). *Ten Great Ideas about Chance* (Princeton University Press, Princeton, NJ).

Dilworth C. (2013). *Simplicity: A Meta-metaphysics* (Lexington Books, Lanham, MD).

Eagle A. (2018). Chance Versus Randomness, *Stanford Encyclopedia of Philosophy*. Available at https://plato.stanford.edu/entries/chance-randomness/.

Gärdenfors P. (1975). Qualitative probability as an intentional logic, *Journal of Philosophical Logic*, 4(2), 171–185.

Gillies D. (2000). *Philosophical Theories of Probability* (Routledge, London).

Giry M. (1982). A categorical approach to probability theory, In B. Banaschewski (ed.), *Categorical Aspects of Topology and Analysis*, Vol. 915 (Springer, Berlin; New York).

Gnedenko B.V. and Kolmogorov A.N. (1954). *Limit Distributions for Sums of Independent Random Variables*; translated by K.L. Chung, Addison-Wesley, Boston, MA.

Griffiths T.L., Daniels D., Austerweil J.L., and Tenenbaumd J.B. (2018). Subjective randomness as statistical inference, *Cognitive Psychology*, 103, 85–109.

Harper D. (2000). *Online Etymology Dictionary*, Available at https://www.etymonline.com/word/probable.

Heunen C., Kammar O., Staton S., and Yang H. (2017). A convenient category for higher order probability theory, arXiv:1701.02547.

Illari P. and Russo F. (2014). *Causality: Philosophical Theory Meets Scientific Practice* (Oxford University Press, Oxford).

Jaynes E.T. (2003). *Probability Theory: The Logic of Science* (Cambridge University Press, Cambridge).

Jeffreys H. (1939). *Theory of Probability* (Oxford University Press, Oxford).

Keynes J.M. (1921). *A Treatise of Probability* (MacMillan & Co., London; New York).

Khrennikov A. (2007). Generalized probabilities taking values in non-Archimedean fields and in topological groups, *Russian Journal of Mathematical Physics*, 14, 142–159.

Kolmogorov A. (1933). *Grundbegriffe der Wahrscheinlichkeitsrechnung* (Springer, Berlin; New York); translated as *Foundations of the Theory of Probability* (Dover Publ. Inc., Chicago), (2018).

Laplace P.S. (1812). *Théorie Analytique des Probabilités* (Courcier, Paris).

Laplace P.S. (1814). *Essai Philosophique des Probabilités* (Courcier, Paris), translated as: *A Philosophical Essay on Probabilities* (1995), (Springer, Berlin; New York).

Lewis D.K. (1976). A subjectivist's guide to objective chance, In R.C. Jeffrey (ed.), *Studies in Inductive Logic and Probability*, Vol. II, 263–293 (University of California Press, Berkeley).

Mann B. (1994). How many times should you shuffle a deck of cards?, *UMAP Journal*, 15(4), 303–331.

Meyer P.A. (1993). *Quantum Probability for Probabilists* (Springer, Berlin; New York).

Mörters P. and Peres Y. (2010). *Brownian Motion* (Cambridge University Press, Cambridge).

Müller T., Rumberg A., and Wagner V. (2019). An introduction to real possibilities, indeterminism, and free will: Three contingencies of the debate, *Synthese*, 196(1), 1–10.

Nickerson R.S. (2002). The production and perception of randomness, *Psychological Review*, 109, 330–357.

Okes D. (2019). *Root Cause Analysis, The Core of Problem Solving and Corrective Action*, 2nd edition (ASQ Quality Press, Milwaukee, WI).

Poincaré H. (1916–1954). *Oeuvres de Henri Poincaré publiées sur les auspices del l'Académie des Science* (Gauthier-Villars, Paris).

Poisson S.D. (1837). *Recherche sur la Probabilité de Jugements en Matière Criminelle et en Matière Civile* (Bachelier, Paris).

Popper K.R. (1959). The propensity interpretation of probability, *The British Journal for the Philosophy of Science*, 10(37), 25–42.

Ramsey F.P. (1931). Truth and probability, In R.B. Braithwaite (ed.), *The Foundations of Mathematics and other Logical Essays* (Kegan Paul, Trench, Trubner & Co., New York).

Robitaille D. (2004). *Root Cause Analysis: Basic Tools and Techniques* (Paton Press, Chico, CA).

Rocchi P. (2014). *Janus-faced Probability* (Springer, Berlin; New York).

Rocchi P., Pandolfi S., and Rocchi L. (2010). Classical and Bayesian statistics: A survey upon the dualist production, *International Journal of Pure and Applied Mathematics*, 58(3), 255–280.

Savage L.J. (1954). *The Foundations of Statistics* (John Wiley & Sons, New York).

Skyrms B. (1980). *Causal Necessity* (Yale University Press, New Haven, CT).

Steinhaus H. (1922). Les probabilités dénombrables et leur rapport à la théorie de la mesure, *Fundamenta Mathematicae*, 4, 286–310.

Stern J. (2011). Spencer-Brown vs. probability and statistics: Entropy's testimony on subjective and objective randomness, *Information*, 2, 277–301.

Teigen K.H. and Keren G. (2020). Are random events perceived as rare? On the relationship between perceived randomness and outcome probability, *Memory and Cognition*, https://psycnet.apa.org/record/2020-00962-001.

Tukey J.W. (1960). Conclusions versus Decisions, *Technometrics*, 2, 424–433.

Venn J. (2018). *The Logic of Chance* (Dover Publications, Chicago).

von Mises R. (1928). *Wahrscheinlichkeit, Statistik und Wahrheit* (Springer, Berlin; New York); translated as *Probability, Statistics and Truth* (1957) (MacMillan Co., London; New York).

Walley P. (1991). *Statistical Reasoning with Imprecise Probabilities* (Chapman and Hall, Boca Raton, FL).

Weisberg J. (2011). Varieties of Bayesianism, In D. Gabbay, S. Hartmann, and J. Woods (eds.), *Handbook for the History of Logic*, Vol. 10, 477–551 (Elsevier, London).

Werndl C. (2009). What are the new implications of chaos for unpredictability?, *The British Journal for the Philosophy of Science*, 60, 195–220.

Xu G. and Xu J. (2013). *Orbits: 2nd Order Singularity-free Solutions* (Springer, Berlin; New York).

Chapter 3

Influential Writers

The preceding chapter points the finger against the 'direct theorization' of probability and I imagine that the reader feels this criticism to be somewhat surprising and perhaps eccentric. The direct theorization method is universally shared by probabilists and, at this point, bibliographical quotations should conscientiously support the conclusions of the previous pages.

3.1 Textual Analysis

Citations and sentences should be brought to attest the simplified logic of probability theorists, but this method might present significant limits. Bibliographic references could provide fragmentary evidence, offer distorted views, and lead to controversial conclusions. Hence, I have decided to apply a different technique. I plan to discuss the works of eight eminent masters on the basis of *textual analysis* which is capable of presenting a more reliable and objective landscape.

Textual analysis is not so popular among mathematicians; to the best of my knowledge this is the first attempt to break down texts on probability into their components.

3.1.1 Textual analysis is not bibliographical analysis. The latter ordinarily consists of examining an ensemble of works dealing with a certain topic. A commentator selects the writings according to his intellectual interest, knowledge, and so on. He may choose the pages in keeping with his opinion and may place discordant sources toward the back of the stage. A bibliographical inquiry could be seen as a survey affected by personal criteria, and the conclusion could raise suspicions of partiality and arbitrariness.

Textual analysis focuses on a predefined set of works and requires the researcher to investigate the objective content of each work closely (Gee 2011). Textual analysis could be defined as a way of turning qualitative data into quantitative data (Frey *et al.*, 1999). For example, a textual analyzer counts the number of times certain phrases or words occur; he defines the linguistic structure of the text, the chapters, and the sections; and he dissects the author's narrative technique, the number of times a specific theme is raised, and so forth.

A researcher could use software tools – such as *text mining* programs (Berry and Kogan, 2010) – which even provide visual models, but these go beyond the scope of this inquiry.

3.1.2 I have chosen the *books* B and the *essay* E (Table 3.1) that more extensively describe the thought of the following writers: John Venn, Richard von Mises, Hans Reichenbach, John Maynard Keynes, Frank Plumpton Ramsey, Bruno de Finetti, Leonard Savage, and Andrei Kolmogorov.

3.1.3 Here, I partition the commentary into two because of the different styles of the writings emerging from the textual analysis. On one side there are the works numbered from #1 to #7 and authored by frequentists, subjectivists, Bayesians, and logicians; on the other side is work #8 by Kolmogorov. The first group is ordered by the number of pages (Column **W**, Table 3.1), while the work of Ramsey ranks at the bottom because it is an essay. Column **V** shows the percentage increase of the book sizes with respect to #8

$$\mathbf{V} = (\mathbf{W} - 84)/84.$$

The right-hand side of Table 3.1 exhibits the numbers of the most significant chapters occupied by criticism against concurrent theories (column **C**), the total extent of these chapters (column **Q**), the proportion of the total number of pages (column **Q/W**), and the schools of thought chiefly criticized by the author (column **K**).

Table 3.1. Textual features of the surveyed works.

#	Work					Criticism			
	Reference	Type	W(Pages)	V %	C	Q (Pages)	Q/W	K
1	Von Mises (1957)	B	245	+191	3	20	8.1	Laplacian Subjectivist
2	Savage (1954)	B	309	+267	4	3	0.97	Frequentist
3	Reichenbach (1949)	B	489	+482	9	21	4.2	Subjectivist
4	Venn (1888)	B	508	+504	6; 10	78	15.3	Subjectivist
5	Keynes (1921)	B	539	+541	7; 8	23	4.2	Frequentist
6	De Finetti (1970)	B	769	+815	1; 6; 12	34	4.4	Frequentist
7	Ramsey (1931)	E	43	N.A.	1; 2	8	18.6	Frequentist Inferential
8	Kolmogorov (1933)	B	84	0	0	0	0	N.A.

3.2 A Pure Mathematician

The textual analysis shows four characteristics exclusive to #8:

1. The book is 84 pages long, which is the minimum size.
2. The book deals only with mathematical themes.
3. The book does not present remarks against alternative theoretical models ($\mathbf{C} = 0$).
4. The book does not contain any argument or philosophical considerations ($\mathbf{Q} = 0$).

3.2.1 These features demonstrate that Kolmogorov has purely mathematical intents. He is cognizant of the sad state of this sector and aims to establish a solid formal basis for probability calculus (Shafer and Vovk, 2018). This formulation is entirely satisfactory from the mathematical viewpoint, problems still remain in several fields and the Soviet scientist is perfectly aware of the limits of his work:

> "Our system of axioms is not, however, complete, for in various problems in the theory of probability different fields of probability have to be examined."

A formal system is complete when it demonstrates any true statement about the topic under investigation and provides the basis for the systematic grasp of whatever is the pressing problem. Kolmogorov honestly recognizes that his construct has not been completely developed because it neglects the themes formulated by frequentists, subjectivists, Bayesians, etc., which are impossible to evade.

3.2.2 But *there is something more which we can comprehend from* the original text:

> "Let E be a collection of elements $\xi, \eta, \zeta, \ldots$ which we shall call *elementary events*, and $\mathcal{F}$ a set of subsets of E; the elements of the set $\mathcal{F}$ will be called *random events*.
>
>   I. *$\mathcal{F}$ is a field of sets.*
>   II. *$\mathcal{F}$ contains the set E.*
> III. *To each set A in $\mathcal{F}$ is assigned a non-negative real number $\mathrm{P}(A)$. This number $\mathrm{P}(A)$ is called probability of the event A.*

IV. P(E) *equals* 1.
V. *If A and B have no element in common, then*

$$P(A + B) = P(A) + P(B)$$

A system of sets, $\mathcal{F}$, together with a definite assignment of numbers P(A), satisfying Axioms I-V, is called a *field of probability*."

Kolmogorov, besides 'probability', uses the notions of 'initial conditions', 'random event', and other concepts that should be detailed in order to express the distinctive traits of this domain of knowledge fully. These notions can be defined 'tacit axioms', since the definitions lack specificity, and some commentators feel the axiomatic system is somewhat 'watery' and bewail the way in which it has been formulated without explicit reference to indeterminism (Bewersdorff, 2005). This defect can be objectively demonstrated by quantities that have nothing to do with P and satisfy the same axioms.

Example. A company has the mission of transporting loads of items by means of standard containers. Every load X has the same volume of the container A

$$V_A = V_X.$$

The company usually employs two or more containers to ship one load because of the irregular shapes of the items belonging to X. The company calculates the *filling rate* $R(A)$ which is the percentage of the volume occupied by the group of k items of X placed inside the container A

$$R(A) = \frac{V_k}{V_A}.$$

The filling rate serves to establish the transport fares; it lays off the probability calculus, yet it obeys the Kolmogorovian postulates. At first, the generic container A may be more or less filled

$$0 \leq R(A) \leq 1.$$

If A_1 and A_2 are two partially occupied containers, then

$$R(A_1 + A_2) = R(A_1) + R(A_2).$$

In consequence of the additivity property, the whole set Ω of the containers used to transport a stock satisfies

$$R(\Omega) = 1.$$

A large part of the scientific community – and me too – admires the rigor of the axiomatic framework developed under the aegis of measure theory, while the aforesaid weak points wait for accurate justifications as we shall see in the second part of the book.

3.3　Thinkers

The textual analysis of works #1-7 presents the following most evident features:

1. The non-trivial number of pages **W** and **V** reveals the verbose style of these works, which are enriched by comments, reflections, notes, explanations, and remarks. Even if an author formulates his construction by means of formal definitions and axioms (e.g., Keynes, von Mises, and Reichenbach), he fills several pages with verbal annotations.

2. Some sections of the texts are labeled as follows: 'A problem of terminology' (#1 p. 93), 'The nihilists' (#1 p. 97), 'The world and the state of the world' (#2 p. 8), 'The value of observation' (#2 p. 125), 'Distinction between logical and psychological view' (#4 p. 129), 'The application of probability to conduct' (#5 p. 351), and 'Tyranny of language' (#6 p. 28). These and other titles exhibit the concern of writers for qualitative and philosophical themes. They argue about a variety of topics that are rather distant from pure mathematics.

3. The writers disseminate critical commentaries against the concurrent schools; they set down quite a number of notes disapproving of the competitors. Columns **C, Q, Q/W,** and **K** give an idea of the authors' polemical efforts, while Appendix A gives a summary of the unfavorable judgments made by each one. It could be said that the appendix gives an orderly expansion of the four right-hand columns of Table 3.1.

4. Each writer analyzes various aspects of indeterminism (points **1** and **2**). He is able to embrace a large area of this domain of knowledge but confronts the definition of probability directly. He focuses on the side of P that he deems to be true and theorizes about it. None of the authors uses any introductory notion to explain P, except for the frequentists who derive the meaning of $P(C)$ from the *collective* C, which however is a special case.

Let us analyze the common behavior of the masters, concisely outlined in point **4**.

3.3.1 *Minimalist philosophy* – We have seen how a mathematician cannot use X as a primary notion if X has multiple and irreconcilable aspects (Section 2.5.3). This is the case of P that cannot play the role of primitive, but each author aims to theorize probability directly and confines himself to one side of P, which turns out to be intuitive and allows for the creation of a coherent theory. In this way, the author who directly addresses P is convinced that he is formulating the dreamed concept of 'true' ('genuine' or 'authentic') probability.

Authors #1–7, who apply this approach, basically adhere to a definite methodological philosophy that can be called 'minimalist' because it relinquishes the complete construction. This intellectual trend can be summarized by the following *principle of uniqueness* (PU)

"One probability interpretation (or two at most) suffices".

Essentially, this methodological guideline drives all the surveyed inquiries; it legitimizes all the partial theoretical results, and disregards the comprehensive understanding of probability (Figure 3.1). PU unites the minds regardless of whether they reach different conclusions.

The common belief that P can be directly theorized, and a sole model is enough, constitutes a sort of 'hidden postulate', because nobody has so far proved it, nor has any writer openly declared his adhesion to this precise methodological mode. Frequentists, subjectivists, logicians, and others do not officially make known

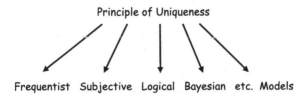

Figure 3.1. Hierarchical conceptual map of the research strategies.

the minimalist philosophy which they embrace with all their soul. Nobody asserts that probability must be directly theorized and reduced to one or two models, and nobody proves or justifies his belief. This tacit common credence turns out to be difficult to contest because it is taken for granted. The different interpretations crash against one another, while PU, which gives rise to the conflicts, is kept away from notice. The scientific community argues about the irreconcilable meanings of P and overlooks the core of the controversies which is concealed from sight.

3.3.2 *Theorists' hypotheses* – Writers #1–7 put the minimalist philosophy into practice by selecting their preferred interpretation of probability. Textual analysis shows how each one develops serious considerations in order to justify his assumption (point **1**). Every theorist sustains his preferred model with all his heart and rejects opposing views. He demolishes concurrent schemes with as much dexterity as he can (point **3**): Venn contests against Laplace; von Mises criticizes subjectivists; Ramsey argues against Keynes; Savage refuses to accept frequentism, and so on. Anyway, all the efforts oriented to destroy the alternative constructions are doomed to defeat since probability has several facets and none is better, or more authentic, or truer than the others.

Another reason frustrates their efforts.

Speaking in general, the conclusion of a theory can be falsified, but the assumptions cannot be disproved or negated. One can erect a construction on the basis of premises which are rational or idealistic, fanciful or unrealistic, provided those premises are effective (Kanazawa, 1998). Paul Feyerabend (1987) holds that anything can lead to formulating the hypotheses of a theory, such as guess, chance,

Table 3.2. Basic choices of the frequentist and subjectivist writers.

	Von Mises	**De Finetti**
What is important	Experimental control of probability	Universal applicability of probability
What does not matter	Universal applicability of probability	Experimental control of probability

intuition, and metaphysical concepts, in fact a hypothetical deductive theory provides adequate explanations that often expect the generation of new ideas.

The author who selects the preferred model of P, factually establishes the assumptions of his theory that derive from personal convictions. Individual judgments appear to be evident when we put two positions side by side. We find out that the feature which has noteworthy importance for an author is negligible for another and vice versa. As an example, let us place the viewpoints of von Mises and de Finetti next to each other. The first assigns superiority to experimental control and does not worry about whether probability applies to the restricted situations that are the collectives. The second wants to use probability calculus in all circumstances and does not care if P cannot be controlled by means of experiments. Von Mises blames the unrealism of the subjective model that de Finetti neglects; in turn de Finetti criticizes the limited coverage of the frequency model which von Mises overlooks (Table 3.2).

As the end result of this situation, the scientific community faces contradictory claims, which nobody can falsify since, by definition, any theoretical hypothesis is legitimate. There is no objective and undisputable criterion by which to establish the superiority of a probability model. The principle of uniqueness has brought experts to an intellectual standstill which they struggle to overcome.

3.3.3 *Philosophy and mathematics* – We cannot contest theoretical hypotheses; however, we can accurately analyze the minimalist philosophy which necessitated the formulation of these hypotheses.

As first, the principle of uniqueness makes the life of the authors uneasy. They are forced to develop extended annotations to show how a single model should suffice. They fill dozens of pages in their strenuous efforts to treat the multi-faceted nature of probability from only one stance (points **1** and **2**).

Despite their laborious attempts, there is no way to substantiate all the tenets, properties, relations, and consequences when one is armed with a partial model of P. The arguments are similar to cages that contain lively animals; the cages eventually break down; the animals escape, and the owner acknowledges that he can no longer control them. In fact, each author ends up recognizing the topics that he cannot treat. Von Mises notes how his theory does not comprehend the single case and concludes: "'Probability of death' when it refers to a single person, has no meaning at all for all of us". De Finetti, conscious that the subjective model has no physical significance, recognizes that "Probability does not exist".

3.3.4 Speaking in general, philosophers provide significant insights and even provide precious support for mathematical studies. They forge and refine new ideas, theories, and perspectives. Thinkers sometimes condense the achievements of different disciplines and improve scholars' awareness. They penetrate mental visions and anticipate scientific discoveries.

The key issue is that a philosophical proposition must correspond to the real world. If and only if the proposition accurately characterizes or describes or matches up with reality, then the philosophical approach is correct and effective. Instead, *the minimalist philosophy* – summarized by PU – *denies the undisputable multifold nature of P,* which came to light in the course of three centuries. It negates the cultural landscape that researchers progressively discovered.

While philosophy normally broadens scholars' vision and the connection between ideas, PU narrows our view about probability and hampers the progress of mathematicians. The minimalist philosophy ignores the prismatic essence of probability and leads to the renunciation of the comprehensive theory, which is expected to explain and unify all the properties of probability calculus.

3.4 The Great Waiver

In conclusion, the textual analysis shows that Venn, Savage, Ramsey, and others are aware of the complex network of tenets underpinning the present sector; however, they try to formulate P *per se* as if it were univocal and self-explanatory, and do not exploit a simpler notion to arrive to probability.

However, this approach cannot be put into effect, it is impossible to carry out, and each author necessarily chooses one model of P (or two at most) in order to create his construct. They put the methodological philosophy, which I label 'minimalist', in practice and consequently the literature presents fragmentary works, each of which casts light on one side of the present domain of knowledge.

For this reason, I do not mean to demolish the importance of each achievement. In point of fact, modern statisticians employ the classical and Bayesian inferences as effective instruments; no scholar disregards the calculus of the favorable and total cases; experimentalists assess frequencies in long-run events, logicians argue about the implications of uncertain propositions, and so forth.

Present day experts accept the diverse interpretations one by one; yet this expedient offers insufficient support (Chapter 5). The core question about the absence of a solid and inclusive framework which harmonizes and integrates all the views remains. The masters who are committed to the 'minimalist' vision have renounced building up the comprehensive edifice hosting all the probability varieties. The waiver of erecting the complete intellectual framework turns out to be the most relevant fact supporting the thesis of this book.

References

Berry M.W. and Kogan J. (2010). *Text Mining: Applications and Theory* (Wiley, New York).

Bewersdorff J. (2005). *Luck, Logic, and White Lies: The Mathematics of Games* (A. K. Peters, Natick, MA).

de Finetti B. (1970). *Teoria della Probabilità* (Einaudi, Torino).

Feyerabend P.K. (1987). *Farewell to Reason* (Verso, London).

Frey L., Botan C., and Kreps G. (1999). *Investigating Communication: An Introduction to Research Methods* (Allyn & Bacon, Boston, MA).

Gee J.P. (2011). *An Introduction to Discourse Analysis: Theory and Method*, 3rd edition (Routledge, London).

Kanazawa S. (1998). In defense of unrealistic assumptions, *Sociological Theory*, 16(2), 193–204.

Keynes J.M. (1921). *A Treatise of Probability* (MacMillan & Co., London; New York).

Kolmogorov A. (1933). *Grundbegriffe der Wahrscheinlichkeitsrechnung* (Springer, Berlin; New York); translated as *Foundations of the Theory of Probability* (2018) (Dover Publ. Inc., Chicago).

Ramsey F.P. (1931). Truth and probability, In R.B. Braithwaite (ed.), *The Foundations of Mathematics and other Logical Essays*, Ch. VII, 156–198 (Kegan, Paul, Trench, Trubner & Co. Ltd., New York).

Reichenbach H. (1949). *The Theory of Probability: An Inquiry into the Logical and Mathematical Foundations of the Calculus of Probability* (University of California Press, Berkeley).

Savage L.J. (1954). *The Foundations of Statistics* (John Wiley & Sons, New York).

Shafer G. and Vovk V. (2018). *The Origins and Legacy of Kolmogorov's Grundbegriffe*, arXiv:1802.06071.

Venn J. (1888). *The Logic of Chance* (MacMillan & Co., London; New York).

von Mises R. (1957). *Probability, Statistics and Truth* (MacMillan Co., London; New York).

Chapter 4

What Probability Assesses

Chapter 2 shows how probability is a multifold concept and raises this inevitable query:

What exactly does probability evaluate?

4.1 The Logics of the Certain and the Uncertain

Let us start from the root of the problem.

For centuries philosophers and scientists adhered to *classical* or *traditional logic,* which accepts only conclusions that are demonstrably necessary or self-evident. For example, they employed syllogistic forms of deduction that do not allow uncertain, vague, or undefined terms. Even if historians have found traces of exceptions to this intellectual trend, those exceptions formed a negligible fraction from the Grecian and Roman periods up to the French Revolution. The philosopher community followed through the apodictic logic without appreciable deviation (Schneider, 1980).

4.1.1 In broad strokes, the classical method of reasoning involves three essential principles:

A. *Two truth values*: A proposition may be either true or false, other values are not allowed.
B. *The principle of non-contradiction*: A proposition has exactly one value because truth and falsity have irreconcilable meanings and the statements A and not-A cannot hold. Two contradictory

propositions cannot both be true in the same sense at the same time.

C. *Causal determinism*: All phenomena involving inanimate matter or conscious beings are determined by previous events. Causal determinism claims that if one knows the state of the universe at any given moment, one can in theory predict the future, including human actions.

A completely different logic underpins the indeterministic reasoning.

a. Multi-valued logic calculates *n truth values* ($n > 2$), namely, a statement may be true or false or uncertain. A statement can have even infinite continuous values varying between truth and falsity. This property proves to be irreconcilable with **A**.

b. Multi-valued logic places *one or more intermediate values between truth and falsity*, and thus the contrary of an assertion is not completely self-contradictory or impossible and point **B** loses significance.

c. Multi-valued logic qualifies outcomes that are not established by precise origin and *the causal determinism* **C** *does not corroborate the conclusions* in a systematic manner.

The features **a**, **b**, and **c** demonstrate how Pascal's project involved a system of logical rules that opposed the past rules. He called for passing from the '*logic of the certain*' to the '*logic of the uncertain*' – as de Finetti (1989) liked to say – which presents diametrically opposed properties. The 'Geometry of Chance' was not merely a mathematical sector to inaugurate as the pioneers believed; it was a route leading to a novel, boundless, and also challenging territory for theorists and thinkers.

4.1.2 With time, scientists passed from arguing about truth to arguing about plausibility, possibility, and chance, and realized how the new questions required them to overturn usual thinking. Laplace (1995) comments on this progressive and unexpected discovery:

"It is remarkable that a science, which began with the consideration of games, should be elevated to the rank of the most important subject of human knowledge."

4.1.3 Traditional logic permeates all the areas of human knowledge and, in parallel, the 'logic of the uncertain' crosses the same areas from the alternative perspective. Alan Hájek (2011) presents a fair summary:

"Probability is virtually ubiquitous. It plays a role in almost all the sciences. It underpins much of the social sciences – witness the prevalent use of statistical testing, confidence intervals, regression methods, and so on. It finds its way, moreover, into much of philosophy. In epistemology, the philosophy of mind, and cognitive science, we see states of opinion being modeled by subjective probability functions, and learning being modeled by the updating of such functions. Since probability theory is central to decision theory and game theory, it has ramifications for ethics and political philosophy. It figures prominently in such staples of metaphysics as causation and laws of nature. It appears again in the philosophy of science in the analysis of confirmation of theories, scientific explanation, and in the philosophy of specific scientific theories, such as quantum mechanics, statistical mechanics, and genetics. It can even take center stage in the philosophy of logic, the philosophy of language, and the philosophy of religion. Thus, problems in the foundations of probability bear at least indirectly, and sometimes directly, upon central scientific, social scientific, and philosophical concerns."

This landscape helps us to sense that, when Pascal inaugurated the probability calculus, he involuntarily ushered in a cultural revolution.

4.1.4 The new branch of mathematics involves innumerable topics that emerge from the immense breath of indeterminism. As the classical logic serves all the sectors of human activities, so 'the logic of the uncertain' and the theory of probability are crucial in any field. They are central everywhere we try to improve knowledge and predictions:

"Statistical and applied probabilistic knowledge is the core of knowledge; statistics is what tells you if something is true, false, or merely anecdotal; it is the "logic of science"; it is the instrument of risk-taking; it is the applied tools of epistemology; you can't be a modern intellectual and not think probabilistically." (Taleb, 2008)

The theory of probability addresses myriads of situations, cognitive elements, objects, devices, possibilities, and so on which challenge the lives of men and women. All this should be sufficient to conclude that a theorist should accurately discuss the argument A in advance of discussing the formulation of $P(A)$.

4.2 A Generic Sentiment

Probabilists should have spent their energies on identifying the entity A that is capable of denoting infinite situations, thoughts, objects, possibilities, and facts.

4.2.1 We have seen how the authors developed partial views (Chapter 3). Each one confines his attention to a rather specific topic, and a concise survey shows:

a. For Laplace, probability calculates *favorable cases*,
b. for Boole, probability qualifies *logical propositions*,
c. for von Mises, a *long sequence of outcomes*,
d. for de Finetti, a *personal belief* regulated by the Dutch book criterion,
e. for Ramsey, a *personal betting quotient* maximizing the expected utility,
f. for Keynes, a *degree of rational belief*,
g. for Savage, *personal credence on the basis of insufficient information*,
h. for Popper, a *physical tendency*, and
i. for Carnap, the *empirical evidence* given for a statement.

This list makes the reductive perspective of the writers more explicit; they overlook the goal of seeking for the universal argument A the necessary requisites to create a complete theory.

4.2.2 I imagine that the present criticism might sound surprising to the reader. The vast majority of experts believe that A has been correctly settled. They are convinced that *the event* or *the result constitutes the universal argument of P.* For the reader this discussion may perhaps not merit any consideration since A has been formally established and universally accepted. My remarks give the impression of being specious and groundless:

Why examine a question which has been definitively resolved?

4.2.3 In reality things are very different.
Since the age of the pioneers, it seemed natural to apply P to what happens or will happen, called event A, or the way in which the event can happen, which is named outcome a. Writers equally attribute probability to the first or the second. For example, we assign 0.5 to "the probability of the flipped coin landing heads ($= A$)" and, more quickly, to "the probability of heads ($= a$)". The two expressions sound tautological.

From the times of Laplace onwards, the concepts of 'case', 'event', and 'result' came to be overlapping and modern books offer descriptions of this kind:

"An event is a set of outcomes" (Nachlas, 2012; Forsyth 2017),
"In the mathematical formulation, the outcomes will be called elementary events." (Hausner, 1977)

There is a dead simple way to define the objects qualified by probability, the writers agree that *the event is a set of outcomes belonging to the event space* (Kolmogorov, 2018)

$$A = \{a\} = \{a_1, a_2, a_3 \ldots\}. \tag{4.1}$$

For probabilists, (4.1) indicates something that simply occurs, and they are not overly concerned with its details and specific properties (Rocchi, 2006). It may be said that probabilists have taken the conceptual equality *event = outcome* for granted so far.

Definition (4.1) gives the impression of accurately fixing the object measured by P, while the items $\mathfrak{a}$, $\mathfrak{b}$, $\mathfrak{c}$, $\mathfrak{d}$ ... $\mathfrak{i}$ guide the calculus from different perspectives. The various viewpoints seem to have a common root and not much distance from one another.

Unfortunately, this generic sentiment has led to unanalyzed understanding; it has given rise to an ensemble of issues, incongruities, genericities, and paradoxes.

4.3 Modern Cognitions of Events

Definition (4.1) states that the event is no more than a *collection of results,* that is to say, A and a are homogeneous in nature. The two entities should have similar characteristics and substance, but all this openly mismatches with scientific knowledge and the ordinary cognition of things.

4.3.1 In English dictionaries the entry for '*event*' refers to *a fact* or *a phenomenon* or *any complex occurrence,* while a '*result*' is understood as something caused or produced by something else. Normally, entity A is a composite process while a is the final consequence or the conclusion of that process, therefore, the event cannot be stated to be a group of upshots. Besides the outcomes, A includes the preliminaries, the components, the boundary conditions, the procedure to follow, the logic which develops the conclusions, etc. The intended system is much more than the ensemble of results, and even the familiar games of chance bring evidence of how the conceptual equality *event* = *results* is misleading.

Example. The following are events:

1. Flipping a coin,
2. Rolling a die,
3. Picking a card from a deck,

which deliver the ensuing outputs in corresponding order:

a. Heads and tails,
b. Numbers from 1 to 6,
c. Cards ranking in diamonds, spades, hearts, or clubs.

Each of the events 1, 2, or 3 is far more than the subsets a, b, or c in corresponding order.

4.3.2 The idea that the results are insufficient to grasp and describe uncertain phenomena is not entirely new:

> "The upshot of this discussion is that chance is a process notion, rather than being entirely determined by features of the outcome to which the surface grammar of chance ascriptions assigns the chance." (Eagle, 2018)

Gnedenko (1967) underscores the importance of the initial circumstances of the aleatory occurrence:

> "An event that may or may not occur when the set of conditions C is realized is called random."

However, these remarks have not so far evolved into a precise formalization; definition (4.1) continues to dominate the field even if it poses communication problems. Carnap and Keynes note ambiguities:

> "The authors who use the term 'event' when they mean kinds of events get into trouble, of course, whenever they want to speak about specific events. The traditional solution is to say 'the happening (or occurrences) of a certain event' instead of 'the event of a certain kind'; sometimes the events are referred to by the term 'single events.'" (Carnap, 1950)

Considering $A = \{a\}$, one could ask:

Is the single event a single set?

Do repeated results make an ensemble of subsets?

And what are the initial conditions of the event?

The queries do not get clear answers since an abyss separates conceptual equality (4.1) from the modern profound cognition of events.

4.3.3 In the first half of the 20th century, thinkers began to argue about events because of the growing use of this notion in science and a reawakening of intellectual interest in the concept of *change* to which the concept of the event seems inextricably tied. The '*theory of event*' is a branch of *analytical philosophy* because events are normally defined as *particulars* that, unlike *universals*, duplicate at different times. Philosophers began by examining the comprehensive description of events and went on to methodically examine their nature, constituents, causality and so on (Lewis, 1986; Quine, 1981; Davidson, 1980).

Just a few quotations are enough to illustrate the pervasiveness of the concept of the event. The *Tractatus Logicus-Philosophicus* (Wittgenstein, 1962) by Ludwig Wittgenstein deals with seven propositions. The 'first proposition' deals with events and begins as follows:

"The world is all that occurs. The world is the totality of events, not of things. The world is determined by the events, and by their being all the events. For the totality of events determines what occurs, and also whatever does not occur. The events in logical space are the world. The world divides into events. Each event can occur or can don't occur while everything else remains the same. That what does occur, an event is the existence of states of the world."

Researchers also became concerned about formal representations. Among the pioneers, Jaegwon Kim (1993) puts forward an expression of this form

$$[(x_1, x_2, \ldots, x_n, t)P^n], \tag{4.2}$$

where $x_1, x_2, \ldots, x_n$ is an ordered n-tuple denoting concrete objects belonging to the event and P^n is the n-adic attribute which specifies the n-tuple at time t. Others, such as Myles Brand and David Lewis, suggest describing events with a structure having a spatial dimension besides the temporal dimension. Philippe Blanchard and Arkadiusz Jadczyk put forward the m-dimensional Hilbert space $\mathscr{A}$ with a fixed basis and the algebra $\mathscr{A}c$ of diagonal matrices to represent events.

Numerous contributions from operations research, computer science, system theory, management science, engineering, and other disciplines put forward accurate views of events and systems (Head, 1967; Hendrik and Schumacher, 1989; Grimmett, 2010) that go far beyond the simplified ideas of probabilists.

4.3.4 In the light of such ample and authoritative studies, the present book adopts the standard terminology and from now onward

(a) The word '*event*' will denote the global occurrence E, while *experiment, trial, case, phenomenon, happening, episode, action,* and *fact* will convey similar significance with different shades.

(b) The '*result*' e of E will be synonymous with the *outcome, upshot,* and *output*.

These precise concepts will guide us to discover incongruities, paradoxes, and misleading ideas.

4.4 Patched Remedies

Let us recall a familiar topic.

4.4.1 One can read in the literature:

> "The essence of probability approach lies in the fact that the event A is the random event <u>under the given condition</u> $\underline{S}$, i.e., the event A may occur or not occur. A characteristic feature of random events is that their regularities can be found only under repeated testing (the condition S). The case in which event A will always occur <u>under the definite condition</u> $\underline{S}$ is called certain event; the case in which event A never occurs <u>under the definite conditions</u> $\underline{S}$ is called impossible event; and the case in which event A may occur <u>under the definite condition</u> S (during testing) is called random." (Tuzlukov 2002) (underlines are mine)

The *initial conditions* S as well as the outcome are essential to recognize the typologies of events but are alien to definition (4.1). Kolmogorov (2018) confines himself to posit the following:

> "There is assumed a complex of conditions $\mathfrak{S}$, which allows for any number of repetitions."

He does not explain analytically what $\mathfrak{S}$ consist of. Basically, the *initial conditions* that support the fundamental concepts of *random, impossible,* and *certain events* are left to common sense (Chapter 3).

4.4.2 Compound events are described by means of the set algebra in consequence of (4.1). The void intersection set characterizes *mutually exclusive* or *incompatible outcomes*

$$\{a1 \cap a2 \cap a3 \cap \ldots \cap am\} = \varnothing \tag{4.3}$$

Compatible outcomes have the intersection set as not empty

$$\{a1 \cap a2 \cap a3 \cap \ldots \cap am\} \neq \varnothing \tag{4.4}$$

It is evident how Eqs. (4.3) and (4.4) do not formulate univocal solutions. More precisely, *they do not express the sufficient conditions for compatible and incompatible results* and an example is enough to prove this.

Example. Take the following random outcomes:

$a_1 = $ It will rain on the 10th of the next month in Rome.

$a_2 = $ The thoroughbred horse K will win the next gallop race in London.

It is readily conspicuous how a_1 and a_2 do not have common elements

$$\{a_1 \cap a_2\} = \varnothing \tag{4.5}$$

However, the empty intersection (4.5) does not guarantee that rain in Rome and the victory of K are mutually exclusive facts; they both can occur by chance.

Definitions (4.3) and (4.4) prove to be insufficient and the authors add up some specifications to repair the theoretical limitations deriving from the set model. They use verbal annotations and explanatory phrases akin to the following:

> "*The results $a_1, a_2, a_3 \ldots am$* are of the same kind."
> "$a_1, a_2, a_3 \ldots am$ are the possible variants of the outcome of the given event."
$$\tag{4.6}$$

Theorists use 'same type', 'variants', 'initial conditions', and other words as *undefined terms*. Instead, verbal clauses cannot substitute mathematical formulations and every notion should be formalized (Section 3.2). In substance, probabilists are obliged to excogitate rough remedies and patched correctives due to the simplified ideas about events and results.

4.5 What Do We Exactly Calculate?

Kolmogorov calculates $P(A_1), P(A_2), P(A_3), \ldots$ where $A_1, A_2, A_3 \ldots$ denote the outcomes, thus we translate these expressions into the current formalism

$$P = P(a_j) \quad j = 1, 2, \ldots \tag{4.7}$$

What are the consequences deriving from (4.1) and (4.7)?

Are they consistent with the calculus operations?

4.5.1 *The classical formula* – When we relate the predicted number of favorable cases to the total cases

$$P = \frac{Q}{N}, \tag{2.1}$$

1] It is evident how N regards the global event

$$N = N(E), \tag{4.8}$$

2] The number of favorable cases necessarily refers to the specific outcome e that pertains to E, hence the numerator of (2.1) is to be written as

$$Q = Q(E, e), \tag{4.9}$$

3] Finally, equiprobability is the prerequisite of (2.1), which implies controlling the overall phenomenon. The problem solver must make the full inventory of E since he cannot conclude that the cases are equally likely on the basis of e.

Putting together remarks 1], 2], and 3] with (2.1), we obtain that the *classical formula* measures *the overall event E issuing its proper outcome*

$$P = \frac{Q(E, \ e)}{N(E)} = P(E, \ e). \tag{4.10}$$

4.5.2 *Distribution functions* – If the outcome of E is the **discrete** random variable x, the *probability mass function* (PMF) assigns P to every x

$$P_E(x) = f_E(x_k) \quad k = 1, 2, 3, \ldots \tag{4.11a}$$

The *probability density function* (PDF) regards the continuous variable x that falls within the particular range generated by the event E. PDF does not furnish the probability of the single value but the probability that x lies in a specific interval

$$P_E(a \le x \le b) = \int_a^b f(x)dx. \tag{4.11b}$$

Both the functions spread through the intended domain of x, it is evident that (4.11a) and (4.11b) give the account of the overall situation

$$P = P_E. \tag{4.11c}$$

4.5.3 Concluding, the classical formula and the probability distributions qualify the result e together with the event E, and demonstrate how the set model (4.1) simplifies things in harmony with the thesis we are discussing in this book.

4.6 Supplementary Information

The reader could object that probabilists do not meet significant troubles in problem solving. After all, experts succeed in calculating the solutions in endless sectors.

Are we looking into a futile topic?

Are we wasting time?

Let us begin with a pair of examples.

Example. What is the probability of getting *an even number* in a game of chance?

The problem solver accomplishes many operations. He ascertains the randomness of the components, the equiprobability of the outputs, all the possible results, the favorable upshots, and analyzes all the sides of the game. At the end of accurate analysis, the outcome – an even number – yields for the event D (flipped die)

$$N(D) = 6;$$

$$Q(D, \ even \ number) = 3;$$

$$P(D, \ even \ number) = 3/6 = 0.5.$$

And for the event R (roulette):

$$N(R) = 37;$$

$$Q(R, \ even \ number) = 18;$$

$$P(R, \ even \ number) = 18/37 = 0.48.$$

Example. The PDF gamma includes the Euler gamma function Γ, the shape parameter α, and the scale parameter β

$$f_W(x) = \frac{\beta^\alpha x^{\alpha-1}}{\Gamma(\alpha)e^{\beta x}}, \quad x, \alpha, \beta > 0.$$

Professionals use the gamma distribution to investigate several real-life fields such as the weather situation. The PDF gamma describes daily precipitation rates over the entire terrestrial globe since empirical data have been stored for decades. Series of 3-monthly, 6-monthly, 12-monthly, and 24-monthly averaged precipitation are gathered for any local geographical area W.

4.6.1 This pair of examples makes it clear that mathematicians can calculate the solutions after learning about the many features of D, R, and W. In practice, they untangle the problems using various informational resources even though definitions (4.1) and (4.7) require mathematicians to consider only the outcomes of the event. They exploit empirical data, unformal annotations, awareness of the circumstances, personal experience, archives, and other sources which play essential roles in getting correct numerical results. Furthermore, an ensemble of instruments, i.e., graphs, diagrams, and other tools provides visual representations of the additional information items (Robertson, 1988).

It is clear that $\{a\}$ and $P(a_j)$ not only do not harm experts, these definitions allow complicated problems to be dealt with in a manageable way. Probabilists can define the distributions of continuous and discrete random variables with ease, they manipulate random vectors, random edges, and other objects typical of the present sector without exertion.

4.6.2 At the end of the previous positive comments, it is necessary to emphasize how *theorizing is not the same as calculating* (Section 2.5). The latter seeks mathematical–numerical solutions, the former pursues logical explanations through inferential reasoning. Problem solvers benefit from extra insights, but theorists cannot do the same. Abundant oral and written information helps practitioners go beyond the limits of (4.1) and (4.7), but theorists are not allowed to play the same trick. A deductive theory begins with rigorous formalisms which verbal comments cannot complete or refine.

A theorist selects the initial notions with great precision since he must deduce all the conclusions from those notions. He must develop correct reasoning, and any defective assumption hinders the whole logic array. Partial definitions hamper the development of correct inferences or anyway inferential theorization advances limping.

Extensive reflection on the deductive and inductive logics typical of the scientific method goes beyond the scope of the present book, and I confine myself to mentioning the contributions supplied by (Suárez, 2004; Pincock, 2012) besides the classical work of Kuhn (1970).

4.7 Problem Solvers are Right, Theorists are Wrong

Assumptions (4.1) and (4.7) do not disturb real-life applications but do not forgive at the theoretical level where the simplified tenets lead to imprecise conclusions and contradictions. Let us discuss the repercussions deriving from the theorists' negligent concern for the probability argument.

4.7.1 Joseph Bertrand poses this paradoxical problem:

Consider an equilateral triangle inscribed in a circle. Suppose a chord of the circle is chosen at random. What is the probability that the chord is longer than a side of the triangle?

The problem statement focuses on the following precise result

$$e = \text{"The chord is longer than a side of the triangle."} \qquad (4.12)$$

The expression $P = P(a_j)$ implies that one probability value must correspond to the outcome (4.12) that is unique; instead, mathematicians provide different values in consequence of the 'ways' the chord may be placed inside the circle (Aerts and Sassoli, 2014).

Gnedenko (1962) notes how the 'ways' are distinct events:

"The three results 'would be appropriate' in three different experiments."

The complete formalism $P = P(E, e)$ makes explicit Gnedenko's mind, so the *random end-point* experiment yields

$$P_1 = P(E_1, e) = 1/3. \tag{4.13}$$

The *random radial point* approach gives

$$P_2 = P(E_2, e) = 1/2. \tag{4.14}$$

The *random midpoint* event leads to

$$P_3 = P(E_3, e) = 1/4. \tag{4.15}$$

Edwin Jaynes (1973) and more recently Marinoff (1994), Wang and Jackson (2011), and others address Bertrand's problem using another method. Jaynes invokes *the principle of maximum ignorance* and assumes the chords are uniformly distributed, and there is no preferred method for placing the chord inside the circle. The fourth experiment, invariant by scale and translation, yields

$$P_4 = P(E_4, e) = 1/2. \tag{4.16}$$

Definition $P = P(a_j)$ postulates a solution conversely, $P = P(E, e)$ enables the accurate formulation of the distinct experiments E_1, E_2, E_3, and E_4, which brings about e. The complete definition (4.10) explains the various solutions, whereas the partial expression (4.7) is not up to the Bertrand paradox.

4.7.2 Martin Gardner (1961) raises the 'boy or girl paradox':
Mr. Smith has two children.
Question # 1: The older child is a girl. What is the probability that both children are girls?
Question # 2: At least one of them is a girl. What is the probability that both children are girls?

Conventional probability theory investigates this unique outcome

$$e = \text{``Both children are girls.''} \tag{4.17}$$

While question #1 presumes the following event:

$$E_1 : (G,G),(G,B).$$

Question #2 focuses on the situation

$$E_2 : (G,G),(G,B),(BB).$$

Assuming equally likely each pair of children, the probability of both children being girls is for question #1

$$P(E_1, e) = 1/2.$$

And for question #2

$$P(E_2, e) = 1/3.$$

Gardner recognizes the second question could yield another probability value since the description of E_2 is ambiguous, e.g.: Children's genders are equally likely? Are there twins? May a child be a transgender?

The 'boy or girl paradox' does not need demanding calculations were it not for the events that $P(a_j)$ and $\{a\}$ partially formalize, instead $P(E, e)$ allows us to specify all the details that pertain to E_1 and E_2.

4.7.3 The current literature presents 4.7.1 and 4.7.2 as '*paradoxes*', literarily these multiple answers are judged as "seemingly absurd or contradictory or ill-founded". In reality, the solutions to the problems are perfect if one adopts the thorough probability argument (E, e). We can reasonably conclude that practitioners work correctly while theorists do not. The latter begin with simplistic definitions and do not sanction the works of the former, which instead are correct. The incomplete formulations (4.1) and (4.7) prevent the solutions from being legitimized and the current theoretical research considers them to be 'paradoxical'.

4.7.4 A group of scholars relates the origin of the 'paradoxes' to '*different event spaces*' or '*a changing event space*'.

These justifications are right but remain on the surface. A mathematician specifies the *space of events* after the accurate examination of E, which is just the topic we are discussing.

This chapter has tried to show how theorists have had little interest in the subject of P to date. Definitions (4.1) and (4.7) show how they simplified or underrated the complicated notions of 'logic of the uncertain' and provide the final support for the thesis of this book.

References

Aerts D. and Sassoli de Bianchi M. (2014). Solving the hard problem of Bertrand's paradox, *Journal of Mathematical Physics*, 55, 083503, arXiv:1403.4139.

Carnap R. (1950). *Logical Foundations of Probability* (University of Chicago Press, Chicago).

Davidson D. (1980). *Essays on Actions and Events* (Oxford University Press, Oxford).

de Finetti (1989). *La Logica dell'Incerto* (Il Saggiatore, Milano).

Eagle A. (2018). Chance versus randomness, *Stanford Encyclopedia of Philosophy*, Available at https://plato.stanford.edu/entries/chance-randomness/.

Forsyth D. (2017). *Probability and Statistics for Computer Science* (Springer, Berlin; New York).

Gardner M. (1961). *The Second Scientific American Book of Mathematical Puzzles and Diversions* (Simon & Schuster, New York).

Getz D. (2007). *Event Studies: Theory, Research and Policy for Planned Events* (Elsevier, London).

Gnedenko B. (1962). *The Theory of Probability* (Chelsea Publishing Co., New York).

Gnedenko B. (1967). *The Theory of Probability* (Chelsea Publishing Co., New York).

Grimmett G. (2010). *Probability on Graphs: Random Processes on Graphs and Lattices* (Cambridge University Press, Cambridge).

Hájek A. (2011). Interpretations of probability, In *Stanford Encyclopedia of Philosophy*, Available at https://plato.stanford.edu/entries/probability-interpret/#CriAdeForIntPro.

Hausner M. (1977). *Elementary Probability Theory* (Springer, Berlin; New York).

Head G.L. (1967). An alternative to defining risk as uncertainty, *Journal of Risk and Insurance*, 34(2), 205–214.

Hendrik N. and Schumacher J.M. (eds.) (1989). *Three Decades of Mathematical System Theory: A Collection of Surveys at the Occasion of the 50th Birthday of Jan C. Willems* (Springer, Berlin; New York).

Jaynes E.T. (1973). The Well-Posed Problem, *Foundations of Physics*, 3, 477–493.

Kim J. and Sosa E. (1993). Causation, nomic subsumption, and the concept of event, In *Supervenience and Mind: Selected Philosophical Essays*, 3–21 (Cambridge University Press, Cambridge).

Kolmogorov A. (2018). *Foundations of the Theory of Probability* (Dover Publ. Co., Chicago).

Kuhn T. (1970). *The Structure of Scientific Revolutions* (University of Chicago Press, Chicago).

Laplace P.S. (1995). *A Philosophical Essay on Probabilities* (Springer, Berlin; New York).

Lewis D. (1986). Events, In *Philosophical Papers*, Vol. 2. (Oxford University Press, Oxford).

Marinoff L. (1994). A resolution of Bertrand's paradox, *Philosophy of Science*, 61, 1–24.

Nachlas J.A. (2012). *Probability Foundations for Engineers* (CRC Press, Boca Raton, FL).

Pincock C. (2012). *Mathematics and Scientific Representation* (Oxford University Press, Oxford).

Quine W.V.O. (1981). Things and their place in theories, In *Theories and Things* (Harvard University Press, Cambridge, MA).

Robertson B. (1988). *How to Draw Charts & Diagrams* (North Light Books, Cincinnati, OH).

Rocchi P. (2006). De Pascal à nos jours: Quelques notes sur l'argument A de la probabilité P(A), *Actes du Congrès Annuel de la Société Canadienne d'Histoire et de Philosophie des Mathématiques*, 19, 228–235.

Schneider I. (1980). Why do we find the origin of a calculus of probabilities in the seventeenth century?, In J. Hintikka, D. Gruender, and E. Agazzi (eds.), *Probabilistic Thinking, Thermodynamics and the Interaction of the History and Philosophy of Science*. Synthese Library, Vol. 146, 3–6 (Springer, Berlin; New York).

Suárez M. (2004). An inferential conception of scientific representation, *Philosophy of Science*, 71(5), 767–779.

Taleb N.N. (2008). *The Fourth Quadrant: A map of the limits of Statistics* (Edge Foundation), Available at https://www.edge.org/conversation/nassim_nicholas_taleb-the-fourth-quadrant-a-map-of-the-limits-of-statistics.

Tuzlukov V. (2002). *Signal Processing Noise*. CRC Press.

Wang J. and Jackson R. (2011). Resolving Bertrand's Probability Paradox, *International Journal of Open Problems in Computer Science and Mathematics*, 4(3), 73–103.

Wittgenstein L. (1962). *Tractatus Logico-philosophicus*; translated by D.F. Pears and B.F. McGuinness (Routledge & Kegan Paul, London).

Chapter 5

Repercussions From the Missing Comprehensive Theory

It seems that Charles Peirce claimed: "In no other branch of mathematics is it so easy for experts to blunder as in probability theory."

5.1 Ambiguities

Actually, the current confusing cultural situation is not limited to abstract disputes; it hinders all those who make forecasts or decisions in everyday life. This non-negligible group of people includes politicians, executives, businessmen, scientists, and also gamblers, bettors, and laymen who remain doubtful about the methods to adopt and the meaning of the numerical results they obtain.

5.1.1 Gábor Székely (1987) authored *Paradoxes in Probability Theory and Mathematical Statistics*, a fine book written for educational scopes which offers an accurate account of nearly 90 problems. Besides the 'official' paradoxes of Stuart, Basu, Friedman, and others, Székely illustrates exercises usually assigned to students and emphasizes how the 'normal' solutions of the exercises present special, inexplicable, or even contradictory aspects. In fact, no question can be assessed on the basis of unified criteria. Even the result of a simple school exercise appears paradoxical in the light of irreconcilable views and divergent philosophical opinions. The previous chapter has offered two exemplifications of problems that plague this area of mathematics in that they are judged to be paradoxical but in fact are not.

5.1.2 Pragmatic thinking has long emphasized the fertility of active knowledge. Several academicians as well as businessmen trust in the leading role of technology and the many ways in which innovation happens. The so-called *utilitarian* writers claim that *technology* drives innovation and jobs, *science* remains in the backstage; for them *theories* lag even further behind. Abstract studies have insignificant impact for those who adhere to utilitarian thinking, and perhaps the reader shares this sentiment.

The following sections mean to recall the far-reaching repercussions of the missing comprehensive theory of probability in two modern fields of activity.

5.2　What is Better?

Nowadays, experts subdivide the *statistical sciences* into *descriptive statistics* and *inferential statistics*. The latter has two main uses: making estimates about a population and testing hypotheses to draw conclusions about a population. However, the remarkable ensemble of methods and practices in use is not univocal, and two schools propose contrasting approaches to inferential statistics.

5.2.1 *Classical* and *Bayesian statistics* agree on the idea that the more information one gets, the more accurate the calculated predictions. Sometimes the two schools adopt symmetrical techniques, but they show evident disparities besides occasional convergence. Listening to Fisher, Neyman, Pearson, Savage, Cox, and others, one hears of very different formulas and objectives.

When the two inferential investigations furnish identical numerical results, they assign irreconcilable meanings to the numbers. Singpurwalla (2002) develops an accurate analysis of this disagreement, which emerges in the reliability sector and directly affects modern industry, commerce, etc.

A manager willing to spend for the optimum statistical inquiry, easily gets different answers from different teams of professionals.

Two statisticians may well disagree about the most suitable formulas for given prerequisites since they adopt incompatible criteria of action. Sometimes they tend to avoid any discussion with the customer who perceives statistics as a religious faith with different beliefs.

5.2.2 Reviewers have analyzed the strengths and weaknesses of each methodology. For example, some point out that p-value and significance level are often subject to misinterpretations in the classical environment; also, experiments must be fully specified ahead of time. Others emphasize there is no precise and univocal way to choose a prior in Bayesian analysis; moreover, Bayesian research sometimes needs such sophisticated computations that it takes much computer time.

Quite a number of studies ponder pros and cons including (Bayarri and Berger, 2004; Efron, 2005; Huber and Train, 2001; Kass, 2011; Sprenger, 2016). Others, such as (Good, 1992; Kim and Schmidt, 2000; Vallverdú, 2016), make attempts to establish an overall evaluation but in reality there is no shared criterion, and no conclusive end point can be reached. Some commentators are inclined to emphasize the distances between the two statistical schools; others combine bootstrapping, choice of estimators, and other specific techniques to minimize the differences extant between the two methodologies.

In conclusion, there is no rigorous criterion to select the best statistical method in a project.

5.2.3 Pragmatic criteria cannot be applied since probability theories back this broad argument, and the sharp opposition among the authors perpetuates the dilemma. Savage (1954) summarizes this deadlock which still persists after decades of intellectual efforts:

> "It is unanimously agreed that statistics depends somehow on probability. But, as to what probability is and how it is connected with statistics, there has seldom been such complete disagreement and breakdown of communication since the Tower of Babel."

5.2.4 Statistics has its early origins in demography. From the 19th century onward, questions of data analysis emerged in assurance, public administration, and science. Today, statistics is invading almost all areas of human work, from engineering to business, from sociology to economics and medicine (Gigerenzer *et al.*, 1989). Experts exploit statistical inference to predict sales, plan political campaigns, determine insurance premiums, organize a medical procedure, forecast market trends, assess risks, design the algorithms of artificial intelligence, and much more. Probability and statistics back top managers who invest billions of dollars, politicians who decide the destiny of nations, and so forth. It is evident how the present issue is not a trivial matter at the practical level, and the second part of the book will put forward an innovative answer.

5.3 The Great Quantum Muddle

Quantum mechanics (QM) presents a large array of processes that seem weird and inexplicable.

5.3.1 The following list briefly recalls some of them:

- Max Planck first discovered that matter and energy cannot be subdivided without any limit. The smallest parts can be emitted or absorbed only in integer multiples of a small unit.
- Photons, electrons, and other tiny elements behave sometimes as waves and sometimes as particles.
- There are experiments in which the wave state of the quantum is fundamentally undetectable because it vanishes.
- Two related variables cannot both be measured exactly. Heisenberg asserts that the more precisely a variable is measured, the less precisely the complementary variable can be known, and vice versa.
- Two particles are able to form an inseparable whole so that they become inextricably linked. Whatever happens to one particle immediately affects the other, regardless of how far apart they are. This strange effect, named *entanglement*, produces a situation in which the behavior of one particle instantaneously determines the state of the other even if this is placed at an enormous distance.

5.3.2 The 1920s and 1930s can be considered as the *Golden Age* of QM, which brought forth significant theoretical achievements

Table 5.1. Leading quantum interpretations.

1	Consistent Histories	9	Quantum Darwinism
2	Copenhagen Interpretation	10	Quantum Information Theories
3	de Broglie–Bohm Theory	11	Quantum Logic
4	Ensemble Interpretation	12	Relational Quantum Mechanics
5	Many Worlds Interpretation	13	Time-Symmetric Quantum Theories
6	Modal Interpretations	14	Transactional Interpretation
7	Objective Collapse Theories	15	Von Neumann–Wigner Interpretation
8	Quantum Bayesianism		

(Jagdish, 2001). Unfortunately, the subsequent production was not up to expectations and several phenomena remain inexplicable to date (Wick, 1994). Researchers have produced a huge amount of work and tons of equations that have perhaps created more questions than answers. They have put forward more than dozen interpretations in order to solve the quantum puzzles (Table 5.1), but there is no consensus on the best framework. Some explanations look rather bizarre if not downright fantastic. Often the interpretation preferred by an expert depends on personal opinions rather than the successful support of experiments (Manzano, 2013).

5.3.3 Where do these difficulties stem from?

Why did brilliant and celebrated scientists miss the target?

The scientific method involves two phases of work that adhere to a strict priority criterion. In the first stage, mathematicians set up the formal instruments; in the second stage, physicists, chemists, and other scientists are able to explain new discoveries employing the instruments that have been prepared in advance. The second cannot solve the problems lacking the support of the first. For example, Hendrik Lorentz perfected the transformation equation, and after a year Einstein published what is now called 'special relativity' using Lorentz's achievements. Einstein could not have formalized his theoretical solution without Lorentz's contribution.

It is reasonable to conclude that the fragmentary theories commented on in this part of the book hinder the progress of quantum physics, which is amply phrased in terms of probability (Beck, 2018; Vervoort, 2012). The comprehensive view of indeterminism, which is still missing, has a direct negative impact on QM, and Karl Popper warmly pinpointed this issue (Jammer, 1991):

> "The interpretation of the formalism of quantum mechanics is closely related to the interpretation of the calculus of probability. (...) I have tried to show for many years, it would be sheer magic if we were able to obtain knowledge – statistical knowledge – out of ignorance (...) as a consequence, we are faced with what I shall call the *great quantum muddle*." (Popper, 1967) (Italics mine)

His essay 'Quantum Mechanics without the Observer' provided the full account of his vision but met with a cold reception (Suppes, 1993). In the abstract, quantum scientists have accepted his rational strategy but, in practice, have followed him little.

For a circle of physicists, probability calculus is correct and works perfectly. They are inclined to believe that the various interpretations basically deal with philosophical themes unrelated to physics. They tend to ignore the declaration of Kolmogorov who openly admits the limits of his construction; they also fail to heed that the collective is a special kind of event and the subjective model is out of control. Khrennikov (2009) remarks:

> "Mathematicians are not interested in quantum physics (mainly because they do not know quantum theory). Physicists are not interested in foundations of probability theory (mainly because they know not so much even about the standard Kolmogorov measure-theoretical approach)"

Karl Popper was convinced – and I altogether share in his conviction – that the diverse viewpoints in QM are fundamentally unsalvageable without the aid of the complete probability theory, and the third part of this book will address this issue.

References

Bayarri M.J. and Berger J.O. (2004). The Interplay of Bayesian and Frequentist Analysis, *Statistical Science*, 19(1), 58–80.

Beck J.L. (2018). Contrasting implications of the frequentist and Bayesian interpretations of probability when applied to quantum mechanics theory, arXiv:1804.02106.

Efron B. (2005). Bayesians, frequentists, and scientists, *Journal of the American Statistical Association*, 10, 1–5.

Gigerenzer G., Swijtink Z., Porter T., and Daston L. (1989). *The Empire of Chance: How Probability Changed Science and Everyday Life* (Cambridge University Press, Cambridge).

Good I.J. (1992). The Bayes/non-Bayes compromise: A brief review, *Journal of the American Statistical Association*, 87(419), 597–606.

Huber J. and Train K.E. (2001). On the similarity of classical and Bayesian estimates of individual mean partworths, *Marketing Letters* 12(3), 259–269.

Jagdish M. (2001). *Golden Age of Theoretical Physics*, Vols. 1–2 (World Scientific, Singapore).

Jammer M. (1991). Sir Karl Popper and his philosophy of physics, *Foundations of Physics*, 21, 1357–1368.

Kass R. (2011). Statistical inference: The big picture, *Statistical Science*, 26(1), 1–9.

Khrennikov A. (2009). *Interpretations of Probability* (De Gruyter, Berlino).

Kim Y. and Schmidt P. (2000). A review and empirical comparison of Bayesian and classical approaches to inference on efficiency levels in stochastic frontier models with panel data, *Journal of Productivity Analysis*, 14, 91–118.

Manzano D. (2013). *The Ongoing Debate About the Foundations of Quantum Mechanics*, Available at https://mappingignorance.org/2013/01/21/the-ongoing-debate-on-the-foundations-of-quantum-mechanics/.

Popper K.R. (1967). Quantum mechanics without "The Observer", In M. Bunge (ed.), *Quantum Theory and Reality. Studies in the Foundations Methodology and Philosophy of Science*, Vol. 2 (Springer, Berlin; New York).

Savage L.J. (1954). *The Foundations of Statistics* (Dover Publications, Chicago).

Singpurwalla N.D. (2002). Some cracks in the empire of chance (flaws in the foundations of reliability), *International Statistical Review*, 70(1), 53–78.

Sprenger J. (2016). Bayesianism vs. frequentism in statistical inference, In A. Hajek and C. Hitchcock (eds.), *Oxford Handbook of Probability and Philosophy* (Oxford University Press, Oxford).

Suppes P. (1993). Popper's analysis of probability in quantum mechanics, In: *Models and Methods in the Philosophy of Science: Selected Essays*, 311–326 (Springer, Berlin; New York).

Székely G.J. (1987). *Paradoxes in Probability Theory and Mathematical Statistics* (Springer, Berlin; New York).

Vallverdú J. (2016). *Bayesians Versus Frequentists: A Philosophical Debate on Statistical Reasoning* (Springer, Berlin; New York).

Vervoort L. (2012). The instrumentalist aspects of quantum mechanics stem from probability theory, *AIP Conference Proceedings*, 1424, 348–354.

Wick D. (1994). *The Infamous Boundary: Seven Decades of Controversy in Quantum Physics* (Birkhauser, Basilea).

Part 2

Toward a Comprehensive Framework

Chapter 6

Discussing a Viable Road

The first part of this book shows how mathematicians underrated the project of Pascal, consequently they developed theories illustrating partial aspects and the dreamed 'Geometry of Chance', namely, the exhaustive probability theory, has not yet come to light. This chapter outlines a possible road to follow in order to get closer to the target.

6.1 A Research Plan

As a logical consequence of the preceding pages, theorists should give up the minimalist philosophy and should seek a unifying and coherent framework that places the many features of P under one roof. They should bring together in one place all the statements that prove to be true.

6.1.1 The new construction should cover the boundless territory of multivalued logic and be abstract enough to include the strikingly different views on P, but concrete enough to find real-world evidence. In this way, the new construction will meet the expectations of academics and practitioners, mathematicians and managers, and even ordinary people dealing with probability and statistics.

6.1.2 A self-explanatory notion is needed to set up the complete construction, but the concept of probability has so many facets that it cannot play this key role. This conclusion — already commented on — is an intellectual breakthrough because it denies the research methods shared by almost all authors.

6.1.3 In summary, the future framework should conform to the following methodological requisites:

(1) Discussing the correct primitives or initial concepts.
(2) Deriving all of the conclusions through inferences.
(3) Verifying the conclusions by means of empirical testing.
(4) Rejecting any personal viewpoint.

A theorist should subsume all knowledge under a coherent logic and fulfill the following tasks:

(a) Demonstrate the various probability interpretations.
(b) Justify probability in relation to practical experience.
(c) Prove the formulas assumed as empirical or postulated so far.
(d) Select the rule for the most appropriate statistics in a project.
(e) Support quantum physics.

Obviously, the intended research project, which centers on the fundamental themes, leaves specialist issues to dedicated works.

6.2 Critical Starting Point

Ordinarily, a hypothetical-deductive theory features a small number of initial propositions and obtains a large set of diverse results through inferential reasoning.

6.2.1 The initial propositions of a mathematical theory present primitives and postulates. The first are self-explanatory, while the postulates illustrate the properties of the primitives or specify or restrict the relations among them.

The majority of probabilists converge on the *non-negativity, normalization,* and *additivity* axioms; however, these axioms deal with complicated concepts. Besides *probability,* I mention *randomness, chance, regularity, causality,* and other collateral notions such as *random events, initial conditions, outcomes,* and so forth. All of them turn out to be anything but simple; therefore, the real challenge does not lie in the axioms (as many believe) but in the notions cited by the three axioms that are not self-evident. The first requisite of Section 6.1 opposes the major obstacle to probability theorization.

6.2.2 *Causal and phenomenological criteria* — A large variety of elements calculated by P have the name '*event*'. Probability theories apply to events, hence, we can conclude that *the event is the cause of probability*. This remark recalls Baruch Spinoza who underscored the role played by *causes* that have the property of originating and sustaining human knowledge (Mignini, 1990; Morfino, 1999):

> "If the thing is not in itself, but requires a cause in order to exist, then its immediate cause must be understood."

> "... a thing is perceived by its essence alone, or by knowledge of its proximate cause."

> "I understand this by its cause whose essence involves its existence."

Spinoza teaches us to look for the proximate cause of probability and we can translate his thought this way: "I understand probability by means of the event whose essence involves the multiform nature of P."

6.2.3 Beyond any doubt, the event determines the numerical value of P, and also determines the meaning of this number. The projected inquiry, which will deduce the base features of $P(E)$ from E, will explore the experiential and lived aspects of phenomena. This method requires exploring how indeterminism is experienced rather than what is thought about those experiences, or the meaning ascribed to them. The planned inquiry will analyze our everyday knowledge while it will suspend the researchers' preconceived assumptions. It may be said that the inquiry will apply a *phenomenological approach* and will develop the inverse intellectual dynamics with respect to the study methods shared by the vast majority of probabilists who start with the preferred model of P.

6.3 Axiomatization

In 1900, David Hilbert made a list of the problems unsolved at that time. In particular, the sixth problem regarded the axiomatization of probability as part of the physical sciences. Kolmogorov (2018) answered Hilbert's plan of research and provided a solid ground for the calculus of probability. It is evident how the phenomenological approach follows another direction:

Does the present plan of study repudiate Hilbert and Kolmogorov?

6.3.1 The construction of the Soviet mathematician towers as a scientific conquest even if some researchers claim his work should be improved (Section 3.2). For example, Burdzy (2009) adds up a special axiom of conditional probability; Rényi (1970) introduces probabilities affected by multiple possible conditions; Knuth and Skilling (2012) intend to integrate with Cox formalisms, and so forth.

The phenomenological approach adopts a different style but rejects neither Hilbert's guidelines nor Kolmogorov's construction. I explain myself using the following bibliographical case.

6.3.2 The literature presents two descriptions of mechanics authored by Newton and Lagrange, respectively (Deriglazov, 2010). They treat the same subject matter, yet the two constructs are very dissimilar. Lagrange (2012) deduces the equations of motion via the use of the *function L* and the *principle of least action*

$$L(q, \dot{q}) = T(q, \dot{q}) - V(q).$$

The kinetic energy T and the potential energy V lead to compact and generalized formulations. For example, conservation of energy arises when L has no explicit time dependence on the basis of this elegant expression

$$\frac{dE}{dt} = -\frac{\partial L}{\partial t}.$$

These and other concise equations support fast calculations and inferences. Lagrange's frame proves to be particularly effective in optimization problems of dynamic systems and problems with holonomic constraints. At the same time the work of the Frenchman exhibits an apparent omission: it does not explain the energies T and V and, in a way, bypasses the basic L function.

6.3.3 I imagine the reader objecting that this is not a real limit; the notion of *force* opens up the study of classical mechanics. *Work* defined as the application of F along with the path leads to *kinetic energy* and *potential energy*, which has a precise form inside the Earth's gravitational field. Newton provides the accurate explanations that Lagrange can evade. This one assumes a view which could be defined as axiomatic since the essential observables are taken as unnecessary for demonstration.

To my perception, something like this should be done in the probability sector.

6.3.4 The expected probability theory is called for to spell out all the conceptual elements which have been bypassed so far as of now (Chapters 2 and 3). The future construct — symmetrical to Newton's work — will supply the analytical accounts that have been overlooked. The axiomatic theory will continue to ensure fast calculations in similitude with the Lagrangian mechanics.

6.3.5 The work of Lagrange did not raise issues thanks to its predecessor. The English and French scientists followed the appropriate time order; instead probabilists did not conform to rational priority. The universally shared introductory formulas guided the Soviet mathematician who devised a nimble formalism that is very effective in problem solving. Therefore, all that remains is returning to the challenging concepts and completing the job.

6.3.6 Joseph-Louis Lagrange did not establish the final theoretical stage. William Hamilton reformulated kinematic properties providing a more abstract understanding of mechanics. His work contributed to setting up statistical and quantum mechanics (Haar, 1971). Later we find the Routh equations (Rumyantsev, 2001), the Jacobi equations, and von Neumann's mechanics (Mauro, 2003) in the literature. This is just to say that once the base concepts consolidate, probability theorists will be able to set up other formal systems. As classical mechanics offered the necessary background for more advanced theorizations, so the comprehensive theory of probability will pave the way to more sophisticated works. The plan of research that I am advocating does not constitute the final target, and will open the door to new mathematical challenges.

References

Burdzy K. (2009). *Search for Certainty: On the Clash of Science and Philosophy of Probability* (World Scientific Pub. Co., Singapore).

Deriglazov A. (2010). *Classical Mechanics: Hamiltonian and Lagrangian Formalism* (Springer, Berlin; New York).

Haar D. (1971). *Elements of Hamiltonian Mechanics* (North-Holland Publishing Company, Amsterdam).

Kolmogorov A. (2018). *Foundations of the Theory of Probability* (Dover Publ. Co., Chicago).

Knuth K.H. and Skilling J. (2012). Foundations of inference, *Axioms*, 1, 38–73, arXiv:1008.4831.

Lagrange J.L. (2012). *Mécanique Analytique* (Nabu Press, Charleston, SC).

Mauro D. (2003). Topics in Koopman-von Neumann theory, arXiv:0301172.

Mignini F. (1990). In order to interpret Spinoza's theory of the third kind of knowledge: Should intuitive science be considered per causam proximam knowledge?, In *Spinoza: Issues and Directions*, Vol. 14, 136–146 (Brill Publ., Leiden).

Morfino V. (1999). L'evoluzione della categoria di causalità in Spinoza, *Rivista di Storia della Filosofia*, 54(2), 239–254.

Rényi A. (1970). *Probability Theory* (American Elsevier Publishing Company, New York).

Rumyantsev V.V. (2001). Routh's equations and variational principles, *Journal of Applied Mathematics and Mechanics*, 65(4), 543–551.

Chapter 7

Elements of a Structural Theory

Since long I have perceived the sad state of the present field and searched for a new framework by the end of the past century (Rocchi, 1999; Rocchi, 2001). Now and then I have developed some parts, so this book collects the scattered sections and provides the full illustration of my theoretical proposal that presents a distinguished trait: it begins with a precise primary notion that is the event and deduces all the results from it.

Obviously, the formulation of the event is to be redone from scratch.

7.1 The Primary Notion

I begin with the dictionaries that define the entry 'event' in a rather simple manner, such as the following (Vocabulary, 2022):

Lexical Entry: Event: *Something that happens or might happen.*

$$(7.1)$$

An event is a happening of any kind: material or mental, simple or intricate, made by living beings or material components, spontaneous or planned, etc. It may be regular or occasional, speedy or slow, certain or impossible. An event may be unique or repeated many times. The literature acknowledges that this broad idea lives up to the large induction base for the 'logic of the uncertain'.

The intuitive notion → The concept → The formal model
of event   of event   of event

Figure 7.1. Intellectual path toward the event modeling.

The expressions of dictionaries fail to provide the precise support necessary for a mathematical edifice, and the reader could object that a spontaneous thought can lead to generic understanding.

This is right. The current chapter begins with the initial notion (7.1) and will attain the precise and formal model of the event through three steps. It will conduct a conceptual analysis that will yield the formalism suitable for the probabilistic perspective (Figure 7.1).

7.1.1 Lexical entry (7.1) intimates that an event is something that ordinarily happens (occurs or exists) within a context. Even the simplest occurrence refers to precise circumstances that are at least characterized by the *initial conditions* and the *results*. The capability of happening implies that there is an *antecedent* (or *forerunner, etc.*) and a *consequent* (*outcome, output,* or *conclusion*). The element that links the antecedent to the consequence is often called a *process, action, operation,* and so forth. In accordance with (7.1), one can conclude that the formal determination of the event is equipped with three principal parts:

(1) The antecedent,
(2) The consequent,
(3) The relation which connects the first to the second.

Elements **1, 2,** and **3** guide the event specification in conformity with probability calculus, which requires the problem solver to be aware of the whole circumstance (Chapter 4) (Rocchi and Burgin, 2020).

7.1.2 The *structure* in the context of Boolean algebra enables the formalization of the elements **1, 2,** and **3** just introduced.

(I) The *basic fundamental triad,* also called *basic named set,* **X** is

$$\mathbf{X} = (X, f, N). \tag{7.2}$$

where

X = the *support* of **X** denoted by S(**X**),

N = the *component of names* (*reflector*) or *set of names* of **X** denoted by $N(\mathbf{X})$,

f = the *naming correspondence of* **X**.

(II) The *Boolean relationships* AND, OR and the *unary relationship* NOT can apply to the triad and its parts.

(III) The binary *conditional relation* (symbolized by the vertical line '|') expresses the influence of an element of the triad on another element, or even the influence of one triad on another.

The three statements provide the base formal support for the present inquiry. Burgin (2011) has elaborated definition **(I)**; Boole introduced the truth operations **(II)**, and the conditional relation **(III)** was first presented by de Moivre (1967). Let us recall the intellectual movements that led to the three theoretical references.

7.1.3 Various forerunners set the track toward the notion of *structure*, among them are Gauss, Galois, and Hilbert. By the 1930s, the French group named 'Nicolas Bourbaki' took the structure as a *collection of sets* with *functions* and *relations* on them and assumed the structures as the roots of mathematics. Bourbaki (1950) identified three *mother structures*: *algebraic*, *topological*, and *order structures*, proving that any other structure can be obtained from one of the mother structures under specific constraints.

The books of Bourbaki attracted the attention of the scientific community even beyond the circle of mathematicians (Corry, 2004). *Structuralism* became a philosophical movement, which influenced sociology, psychology, anthropology, and linguistics in a special manner. This line of thought holds that complicated entities such as society, the mind, and language can be broken down into components whose interrelations create a broad and overarching loom. Structuralism emphasizes the importance of invariant basic patterns and their relationships. Piaget, Saussure, and Levy-Strauss are regarded as eminent members of this intellectual orientation.

The impact of Bourbaki's books reached the apex during the 1960s, but that cultural influence progressively waned due to a variety of factors. For example, some of the abstractions proved to be less useful than initially expected. Notions such as the categories, that are now considered important, were ignored by the Bourbaki group. Other topics remained on the backstage, such as problem solving, algorithms, and combinatorics.

Some mathematicians put forward original contributions for structures. Mark Burgin can be associated to this group of researchers since functions, polynomials, graphs, morphisms, fuzzy sets, multisets, and others are special *named sets*. Burgin began his investigations about unified foundations of mathematics in the late 1980s (1990) and subsequently expanded on them (2004, 2011, 2012).

7.1.4 George Boole, who was convinced that "the subject of probabilities belongs equally to the science of Number and to that of Logic", explored the close connections between the two fields. Specifically, his algebra adopted the symbols x, y, z, ... to represent propositions called *classes*, while the symbols $+$, $-$, $=$, $\&$ ... represented the *operations* or *relations* among the classes that are subject to the rule of interpretation of true/false. Currently, Boolean algebra is popular for the support it provides to the study of digital systems, but factually, nearly a third of the treatise (2009) pursues the goal of clarifying the 'doctrine of probabilities'. He discovered the inequality, also known as the *union bound*, which establishes the upper bound on the probability of occurrence of at least one of a countable number of events in terms of the individual probabilities of the events.

Boole's book did not enjoy popularity in his time. Later it attracted the attention of John Venn who believed that symbolic logic (a term coined by him) was consistent with commonsense reasoning. Charles Peirce's (1967) earliest published work – *On an Improvement in Boole's Calculus of Logic* – delved into the application of Boole's system to probability calculations. Various coeval authors continue to argue about the nexus between probability and logic. Richard T. Cox (1961) shows that by employing symbolic logic it is possible to deduce the rules of probability from two quite primitive notions. Cox's theorem proves that every true–false logic under uncertainty is isomorphic to conditional probability theory from the subjective perspective. He maintains that probability theory is the only theory of inductive inference that abides by logical consistency. Among the most recent elaboration, we find Edwin Jaynes (2003), who sets up Bayesian probability as a multivalued logic for plausible quantitative reasoning, and Hailperin (1986), who demonstrates how the heavy computation of Boolean probabilistic intervals can be quickly solved by using a linear programming approach, such as parametric and integer-mixed programming.

7.1.5 The antagonistic concepts of conditioning and independence are central in probability calculus and it seems they appeared first during the 18th century. In 'The Doctrine of Chances' by de Moivre (1967), we read:

> "Two Events are independent, when they have no connexion one with the other, and that the happening of one neither forwards nor obstructs the happening of the other. Two Events are dependent, when they are so connected together as that the Probability of either's happening is altered by the happening of the other."

De Moivre also gave an equation to calculate conditional probability. In the same period of time, Bayes considered the *inverse probability problem* which requires the use of *conditional probability*. His posthumous essay articulates the idea of computing the probability of a hypothesis given evidence. Bayes's findings did not reach a wide audience until Laplace took up similar problems independently from him a few years later. The French author wrote a presentation of inverse probability and its application to both binomial and location parameter estimation.

Three-valued logic was originally defined by Jan Łukasiewicz in the early 20th century (Marra, 2013), and his school also investigated *conditional events* (Goodman *et al.*, 1991). Modern writers argue about dependent and independent variables. We obtain intriguing contributions from the theory of functions to multivariate statistics, from the theory of process to statistical testing, and so forth (Anderson and Belnap, 1975; van Benthem, 1984). Chapter 15 of this book will go back to this theme.

7.1.6 The present work aims to apply the phenomenological approach. It develops a 'narrative' similar to classical mechanics, in the sense that it will employ the concepts **(I)**, **(II)**, and **(III)** to describe the phenomena typical of all things that happen or are able to happen.

As first, (7.2) supplies the formal model illustrating the generic occurrence.

Definition 7.1. *The event is the triad or structure* **E** *equipped with the antecedent i, the consequent e, and the relation r, as follows:*

$$\mathbf{E} = (i, r, e). \tag{7.3}$$

7.1.7 A broad assortment of applied projects and theoretical inquiries driven by management science, cybernetics, electronics, operational research, etc. adopt the triadic idea. The *input–process–output* (IPO) paradigm was introduced in electrical design by Mealy and Moore around the mid-1950s (Shiva, 1998). Later the IPO scheme migrated into software engineering (Coleman, 2012) and expanded into an assortment of contexts: from psychology (Ilgen *et al.*, 2005) to education (Osman, 1973), from industry (Steiner, 1972) to biology (Steven, 2013) and environmental science (Hawkins *et al.*, 2007). Lastly, I recall the *input–output model* by Leontiev (1986), which represents the interdependencies among different regional economies.

7.1.8 The key role of the structure **E** induced me to label the present proposal as '*structural theory*' (Rocchi, 1998), it will also be called '*comprehensive*' or '*ample theory*' in this book.

7.2 How the Event Takes Place

The event is something that happens or might happen. This peculiar property can be expressed even using the verbs 'to occur', 'to function', 'to be present', 'to exist', 'to come about', etc. The capability of occurring is the essential feature of **E** and needs to be unfolded in the very first step.

Burgin underscores that *a triad is not a triple* since the triple is any set including three elements; instead *'the triad is a system of three connected components'*, in particular, the correspondence f links X with N. Symmetrically, the correspondence r *establishes* the connection between i and e; it may be said that r *maps* the elements of the first subset to those of the second subset.

We can conclude that the relation r *'joins'* i with e in (7.3), and at the same time we say the two elements are *'joined'* by r. The internal components have complementary qualities, thence r can be recognized as the 'active' element of the triad, while i and e are the 'passive' ones. When r links the antecedent with the conclusion, then $\mathbf{E}$ comes true. When r does not bring forth the conclusion e, then $\mathbf{E}$ does not function, and we can reasonably state:

Property 7.1. *Property of the elementary event: The process r, joining the antecedent i with the consequence e, makes the event $\mathbf{E}$ occur.*

$$(7.4)$$

The specialized roles of i, r, and e make the essence of events explicit. In particular, the event perfectly and completely happens if it begins and finishes, and it finishes when the outcome is brought about; $\mathbf{E}$ takes place if and only if e occurs, and it is natural to conclude that there is a biunivocal relation between the existence of the overall event and its final part given the initial conditions i.

Property 7.2. *Property of biunivocal occurrence*

$$\mathbf{E} \Leftrightarrow e \qquad (7.5)$$

Expressions (7.3), (7.4), and (7.5) will play a central role in the coming chapters, which will deduce definitions, properties, theorems, and corollaries following 'operational' logic. This work will draw conclusions from the phenomenological aspects which the structures relate and not from the abstract properties of structures.

Property 7.2 is of great importance even though sometimes it cannot be controlled. Some events – say silent diseases, conspiracies, and covert operations – do not produce visible results.

Most likely this method of study will be perceived as unusual by the reader, so some synopses point out the main conclusions, step by step.

Synopsis

a. The elementary event **E** has the property of occurring and does so by means of the specialized functions of i, r and e.

References

Anderson A.R. and Belnap N.D. (1975). *Entailment: The Logic of Relevance and Necessity* (Princeton University Press, Princeton, NJ).

Boole G. (2009). *An Investigation of the Laws of Thought on Which are Founded the Mathematical Theories of Logic and Probabilities* (Cambridge University Press, Cambridge).

Bourbaki N. (1950). The architecture of mathematics, *American Mathematical Monthly*, 67, 221–232.

Burgin M. (1990). Theory of named sets as a foundational basis for mathematics, *Structures in Mathematical Theories*, San Sebastian, 417–420.

Burgin M. (2004). Named set theory axiomatization: T Theory, *Science Direct Working Paper No* S1574-0358(04)70044-1, Available at https://ssrn.com/abstract=3177551.

Burgin M. (2011). *Theory of Named Sets* (Nova Science Publ., Hauppauge, NY).

Burgin M. (2012). *Structural Reality* (Nova Science Publ., Hauppauge, NY).

Coleman D. (2012). *A Structured Programming Approach to Data* (Springer Verlag, Berlin; New York).

Corry L. (2004). *Modern algebra and the rise of mathematical structures* (Birkhäuser, Basilea).

Cox R.T. (1961). *The Algebra of Probable Inference* (Johns Hopkins University Press, Baltimore, MD).

de Moivre A. (1967). *The Doctrine of Chances or a Method of Calculating the Probabilities of Events in Play* (Chelsea Publ., New York).

Goodman I., Nguyen H., and Walker E. (1991). *Conditional Inference and Logic for Intelligent Systems: A Theory of Measure-free Conditioning* (North Holland, Amsterdam).

Hailperin T. (1986). *Boole's Logic and Probability: A Critical Exposition from the Standpoint of the y Algebra, Logic, and Probability Theory* (North-Holland Publishing Co., Amsterdam).

Hawkins T., Hendrickson C., Higgins C., Matthews H.S., and Sangwon S. (2007). A mixed-unit input-output model for environmental life-cycle assessment and material flow analysis, *Environmental Science and Technology*, 41, 1024–1031.

Ilgen D.R., Hollenbeck J.R., Johnson M., and Jundt D. (2005). Teams in organizations: From input-process-output models to IMOI models, *Annual Review of Psychology*, 56, 517–543.

Jaynes E.T. (2003). *Probability Theory: The Logic of Science* (Cambridge University Press, Cambridge).

Leontief W. (1986). *Input Output Economics* (Oxford University Press, Oxford).

Marra V. (2013). Łukasiewicz logic: An introduction, In G. Bezhanishvili, S. Löbner, V. Marra, and F. Richter (eds.), *Logic, Language, and Computation*, Lecture Notes in Computer Science, Vol. 7758 (Springer, Berlin; New York).

Osman A.C. (1973). A cybernetics paradigm for research and development inaccountancy education and training, *The Vocational Aspect of Education*, 25(62), 105–110.

Peirce C.S. (1967). *Collected Papers of Charles Sanders Peirce*, In C. Harshorne and P. Weiss (eds.) (Oxford University Press, Oxford).

Rocchi P. (1998). *La Probabilità è Obiettiva o Soggettiva?* (Pitagora Editore), translated as: *The Structural Theory of Probability; New Ideas from Computer Science on the Ancient Problem of Probability Interpretation* (2003) (Kluwer/Plenum, New York).

Rocchi P. (1999). Si puo' davvero pensare che l'evento aleatorio sia un concetto trascurabile?, *Induzioni*, 18, 85–93.

Rocchi P. (2001). Toward the accurate model of random events, *Revista de Estatística*, Vol. II, 2, 353–354.

Rocchi P. and Burgin M. (2020). An essay on the prerequisites for the probability theory, *Advances in Pure Mathematics*, special issue on *Probability and Mathematical Statistics*, 10, 685–698.

Shiva S.G. (1998). *Introduction to Logic Design*, 2nd edition (Marcel Dekker, New York).

Steiner I.D. (1972). *Group Process and Productivity* (Academic Press, Cambridge, MA).

Steven F. (2013). Input-output relations in biological systems: Measurement, information and the Hill equation, *Biology Direct*, 8(1), 31.

van Benthem J. (1984). Foundations of Conditional Logic, *Journal of Philosophical Logic*, 13(3), 303–349.

Vocabulary.com (2022). Available at https://www.vocabulary.com/.

Chapter 8

Fields of Application

Probability applies to material and abstract problems which require specific approaches. So far, these approaches have seemed impossible to bring into harmony, even conflicting. This chapter has the purpose of accomplishing just this.

8.1 Material and Mental Events

The 'logic of the uncertain' covers an immense field of applications and we can subdivide this field into two large domains in keeping with the double nature of the event.

8.1.1 The triad $\mathbf{E}$ proves to be self-explanatory *in the physical field* because the structural elements are perceived by the five senses directly or using measurement devices. Experts easily recognize the components of $\mathbf{E}$, and an example of structural analysis should be enough.

Example. The hand flipping the coin from the thumb to the forefinger is the preliminary part of the game (Figure 8.1). The flight r of the coin connects the antecedent to the outcome which is the upward side of the coin

$$\mathbf{E} = (\textit{Flipping, Flight, One face}).$$

8.1.2 The second domain of application assesses human thoughts which are invisible happenings. Reasoning, judgments, credence,

Figure 8.1. The structural components of the heads or tails game.

deductions and so on begin with some premises i and close with a conclusion or a spectrum of conclusions e. When the mental process is correct, r connects i with e and the event comes about. When the reasoning is absurd or senseless, there is no logical link between i and e, hence **E** turns out to be an impossible mental event. In short, human thoughts are occurrences that satisfy Property 7.1.

Logicians, subjectivists, Bayesians and others teach us to describe a reasoning process by means of sentences which make explicit the terms of the problem and are subject to true/false judgements. The propositions, which instantiate the mental elements i, r and e, are printed with normal fonts here. The components of physical occurrences are printed with italic fonts.

8.1.3 Let us analyze some cases.

John Keynes relates probability to the logical thought of the individual. The human mind formulates the appropriate rational judgement on the basis of the initial statement and following *rational inference* or *rational belief* (Aldrich, 2008). The triad represents the mental process that begins with a prelude, true or hypothetically true, and leads to the conclusion

$$\mathbf{E} = (\text{Preliminaries, Rational inference, Conclusion}). \qquad (8.1)$$

Typically, a theorem requires rational belief since it develops a proof starting from precise hypotheses

$$\mathbf{E} = (\text{Hypotheses, Demonstration, Theorem statement}).$$

Carnap (1945) explores inductive reasoning which leads to the closing sentence on the basis of empirical evidence. This triad illustrates the Carnapian view:

$$\mathbf{E} = (\text{Empirical evidence, Logical induction, Conclusion}). \quad (8.2)$$

The following structure should be self-explanatory for the reader:

$\mathbf{E} = ($"I loaned my friend 50€ last February and he failed to pay
me back. I loaned him another 50€ just before Easter.
I solicited him who has not returned so far",
Logical induction, "He will not pay me back").

The subjectivists and the Bayesians assess the following triad where the personal 'Belief' given to the 'Proposition' is established on the basis of initial knowledge (Joice, 2011):

$$\mathbf{E} = (\text{Prior information, Belief, Proposition}). \quad (8.3)$$

For example, the stockholder of the company XY makes the following prevision:

$\mathbf{E} = ($"XY's board of directors presented a brilliant, consolidated
balance sheet", Belief, "I will make money from my shares in a
short while").

The theory of information lies in a rather primitive stage (Rocchi and Resca, 2018). Dozens of ideas have been put forward, but no definition reached the universal consensus so far, hence it is necessary to point out that the clause 'prior information' in (8.3) has generical meaning. It indicates knowledge, experience, cognition, awareness, experience, data, signals, insights and so forth. Chapter 12 will return to this subject again.

8.2 Determinate and Indeterminate Occurrences

Property 7.1 has been established in the abstract and can be fully or partially applied in the world. Scholars and laymen agree that the event can unfold perfectly or even can fail.

8.2.1 The elements i, r and e make the event take place; so, they are used to describe the three types of events presented in the literature. The output e will be shown in the upper right-hand corner of the symbol **E** due to Property 7.2.

Definition 8.1. *The event is certain when r makes a firm connection between the antecedents and the outcome. A dash marks this configuration*

$$\mathbf{E} = (i, r - e) = \mathbf{E}^{[e]}. \tag{8.4}$$

Example. An urn contains only red marbles; a marble is drawn at random and is red. The antecedent consists of the urn with red marbles and the drawing process necessarily brings forth the output

$$\mathbf{E}^{[r]} = (\textit{Urn with red marbles, Extraction} - \textit{One red marble}).$$

Example. The syllogism is a kind of logical deduction that begins with two or more propositions that are stated to be true and arrives at the conclusion that is certain

$$\mathbf{E}^{[Gm]} = (\text{``All men are mortal'' AND ``Greeks are men'',}$$
$$\text{Logical deduction} - \text{``All Greeks are mortal''}).$$

Definition 8.2. *The antecedent never furnishes the outcome of the impossible event. Two vertical lines mark the disagreement between the antecedent and consequent*

$$\mathbf{E} = (i, r \| e) = \mathbf{E}^{\{e\}}. \tag{8.5}$$

Example. An urn contains only red marbles, and an operator draws a black marble. This triad formalizes the impossible consequence of the initial conditions

$$\mathbf{E}^{\{b\}} = (\textit{Urn with red marbles, Extraction} \| \textit{One black marble}).$$

Example. Take the numerical statement: $2 + 3 = 59$. The following structure formalizes the mental event that is impossible to happen:

$$\mathbf{E}^{\{59\}} = ((2, 3), \text{Addition} \| 59).$$

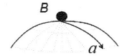

Figure 8.2. Falling ball.

In summary, the certain event surely occurs, the impossible event occurs neither in the world nor in the human mind.

8.2.2 Authors agree that the event, which under a complex of initial conditions sometimes occur and sometimes does not occur, is called random. The outcome *is not entirely determined by the initial elements* in the intermediate case placed between (8.4) and (8.5). Given i and r, the result may or may not occur.

Definition 8.3. *The connection between the antecedents and the outcome is not systematic in the indeterministic event. A semicolon marks the random (aleatory, chancy or uncertain) configuration*

$$\mathbf{E}^{(e)} = (i, r; e). \tag{8.6}$$

Example. The ball B placed at the top of the dome is left free to fall (Figure 8.2). The ball can take any direction, and the initial position i does not ensure the ball will run along the line a (Norton, 2008). Now and then, the ball follows a, the remaining times the ball follows other trajectories

$$\mathbf{E}^{(a)} = (\boldsymbol{B} \; on \; top, \; r \; ; a).$$

Keynes and Carnap conceive of 'partial implications' (Franklin, 2016) as typical of uncertain reasonings. *Proposition i gives some positive evidence for the conclusion but there is no firm connection with e, and the mental event is aleatory.*

Example. Tom notices: "The weather is cloudy, maybe it will rain soon." The inference of the hesitant observer does not guarantee the conclusion. Tom does not relate the premise to the prevision in a definitive manner and the following triad illustrates the uncertain mental occurrence

$$\mathbf{E}^{(rain)} = (\text{"The weather is cloudy"}, \text{Inference}; \text{"It will rain soon"}).$$

Example. As a logical conclusion to the oddly high amount of irid-ium in the Cretaceous–Paleogene strata, it is probable that a meteor caused the extinction of dinosaurs. This remark leads to the very likely but not certain determination of the dinosaurs' end

$$\mathbf{E}^{(extinction)} = (\text{``High amount of iridium''}, \text{Logical induction};$$
$$\text{``A meteor caused the extinction of dinosaurs''}).$$

8.2.3 The definitions just set up show how the outcome is not suf-ficient to determine the type of event, but all the structural compo-nents i, r and e are necessary to establish the classification in explicit terms. In other words, the determinate and indeterminate states are structural properties.

Definitions 8.1 and 8.2 make explicit the *necessary and sufficient conditions* for deterministic events. Definition 8.3 can establish only *the necessary conditions for random events* because physical and mental events present far different characteristics, so the following section and Section 12.7 provide more insights.

Synopsis

a. The elementary event **E** has the property of occurring and does so by means of the specialized functions of i, r and e.
b. Events achieve three main degrees of action and cognition.

8.3 Distinct Perspectives

Material and psychic domains of application necessarily involve log-ics, terminologies and conceptions, which are so different that we recognize distinct perspectives and approaches.

8.3.1 Issues have very different natures and qualities. On the one hand, experts measure the physical tendency of something to occur (Gillies, 2016); on the other hand, they assess how strongly one believes something will come about (Climenhaga, 2020). In a nut-shell, the first approach answers questions of this kind:

How does **E** work?

The second approach focuses on human knowledge and answers questions such as the following:

What do we know about **E**?

The literature frequently calls the distinct domains as *ontological* (also *ontic* or *physical*) and *epistemic* (or *cognitive*). This special jargon is not so ancient, and it is necessary to append how ontology (Alston, 1958) and epistemology (Niiniluoto *et al.*, 2004) are philosophical doctrines originating from cultural environments that were very different from the present context.

8.3.2 The expressions and terminologies employed by experts cast light on the two perspectives which dominate the probability sector (Table 8.1).

The physical stance pays attention to the material elements and processes. The epistemic authors express a person's credence in a given description of a real or imagined circumstance. If initial information is insufficient or inadequate, the individual remains *uncertain* about the conclusion. Instead, on the basis of complete prior information i, an individual becomes aware that e is 'true' or otherwise is 'false'. Table 8.2 makes a list of the different adjectives which authors frequently use in the two contexts.

8.3.3 Rivers of ink have been consumed on paper in order to discuss the two approaches that formally do not have anything in common. Endless debates have been feed to demonstrate the superiority of one mode over the other, while the present phenomenological approach incorporates both the physical and cognitive viewpoints and enables

Table 8.1. The degrees of existence of an event.

Event	Ontological Perspective	Epistemic Perspective
	Given the material premise...	*Given prior information...*
$E^{[e]}$	...**E** *systematically occurs in the world.*	...*the involved individual is sure of* **E**.
$E^{\{e\}}$	...**E** *never occurs in the world.*	...*the individual negates* **E**.
$E^{(e)}$	...**E** *sometimes happens and sometimes does not.*	...*the individual is uncertain about* **E**.

Table 8.2. The attributes of the event and the outcome.

Event	Ontological Attributes	Epistemic Attributes
$E^{[e]}$	*Certain, deterministic, and determinate.*	*True, sure and predictable.*
$E^{\{e\}}$	*Impossible, deterministic, and determinate.*	*False, impossible and predictable.*
$E^{(e)}$	*Random, indeterministic, and indeterminate.*	*Uncertain, unsure and unpredictable.*

us to use very different languages. The concept of structure subsumes all kinds of occurrences and backs probabilists everywhere.

At the same time, the structural analysis keeps the typologies of events rigidly separate and disregards the attempts to superpose or bring them closer. Writers who have this intent, use confusing expressions and often make the issues more complicated than they really are.

Example. "What is the chance that a group of patients found to have the symptom Z actually have the disease J, assuming you know nothing about the persons' symptom Q?"

When we analyze this problem, we find that the symptoms and the disease are objective entities, but the clause "you know nothing about" hints at the idea that the problem has a cognitive nature. The problem solver must necessarily clarify whether the question has an ontological or epistemic nature in advance of starting the calculations.

It must always be clear whether the problem statement raises a material or logical issue, otherwise an ontic answer to an epistemic question (or vice versa) results in a blatant category mistake.

A sentence makes a *category mistake* when it is grammatically well formed, but semantically nonsensical, for example: "The theory of vectors is driving a car" and "Colorless green ideas sleep furiously." The following sentence, merging epistemic and ontic notions, could sound meaningless: "A group of patients has the subjective probability of having caught the disease J." Chapter 12 will provide further insights.

8.3.4 The concept of structure covers all types of occurrences and also supports statisticians who ordinarily focuses on e and formally overlooks $\mathbf{E}^{(e)}$. The triad helps experts to specify the nature of the investigated phenomenon.

Example. A team is investigating the mortality of the country A using the crude death rate (cdr), the overall event has the following structure:

$$\mathbf{E}^{(cdr)} = (A, \ mortality; \ cdr)$$

Suppose the team goes through the hazard rate (hr) of smoking in A. Two alternative phenomena can be formalized this way

$$\mathbf{E}^{(Ahr)} = (A + smoking, \ mortality; \ hr)$$
$$\mathbf{E}^{(Ahr)} = (A + no\text{-}smoking, \ mortality; \ hr)$$

The structure specifies the research hypotheses (Firebaugh, 2008), because statistics does not calculate a bare numerical result but explores the large context $\mathbf{E}^{(e)}$.

References

Aldrich J. (2008). Keynes among the statisticians, *History of Political Economy*, 40(2), 265–316.

Alston W.P. (1958). Ontological commitments, *Philosophical Studies*, 9, 8–17.

Carnap R. (1945). On inductive logic, *Philosophy of Science*, 12(2), 72–97.

Climenhaga N. (2020). The structure of epistemic probabilities, *Philosophical Studies*, 177(2), 1–30.

Firebaugh G. (2008). *Seven Rules for Social Research* (Princeton University Press, Princeton, NJ).

Franklin J. (2016). Logical probability and the strength of mathematical conjectures, *Mathematical Intelligencer*, 38(3), 4–19.

Gillies D. (2016). The propensity interpretation. In A. Hájek and C. Hitchcock (eds.), *Oxford Handbook of Probability and Philosophy*, 406–422 (Oxford University Press, Oxford).

Joice J.M. (2011). The development of subjective Bayesianism, *Handbook of the History of Logic*, 10, 415–475.

Niiniluoto I., Sintonen M., and Wolenski J. (eds.) (2004). *Handbook of Epistemology* (Springer, Berlin; New York).

Norton J. (2008). The dome: An unexpectedly simple failure of determinism, *Philosophy of Science*, 75(5), 786–798.

Rocchi P. and Resca A. (2018). The creativity of authors in defining the concept of information, *Journal of Documentation*, 74(5), 1074–1103.

Chapter 9

Completing the Description of Events

Definition 8.3 says that the expected output is sometimes missing but does not say what happens when the *elementary event* fails. The triad $\mathbf{E}^{(e)} = (i, r; e)$ ignores the alternative outcomes to e.

This chapter addresses this issue and will provide the full account of random events. Dual explanations will be used to make the physical and cognitive perspectives explicit.

9.1 Composite Events

Let us begin with the following concept.

Definition 9.1. *The compound or composite structure $\hat{\mathbf{E}}$ includes m independent sub-structures or sub-events*

$$\hat{\mathbf{E}} = (\mathbf{E}_1, \mathbf{E}_2, \mathbf{E}_3, \ldots \mathbf{E}_m), \quad m \geq 2. \qquad (9.1)$$

In principle, also the sub-events may be compound but, for the sake of simplicity, we confine our attention to the elementary ones.

9.1.1 Property 7.1 makes explicit that the elementary event happens because of the components i, r, and e. Symmetrically, the compound event $\hat{\mathbf{E}}$ happens because of the 'active' components that are the sub-events.

Property 9.1 *Property of the composite event: The sub-events* $F_1, F_2, \ldots, F_m$ *make the composite event* $\hat{E}$ *occur.* $\qquad$ (9.2)

9.1.2 The triad can form a Boolean algebra (Section 7.1.2). In particular, the relationship OR describes the *disjoint event* whose material sub-events operate one by one and make $\hat{E}$ happen in this specific way. In the epistemic domain, the overall judgement is given by propositions expressing mutually exclusive truths.

Definition 9.2. *The disjoint event has this structure*

$$\hat{E} = (E_1 \text{ OR } E_2 \text{ OR } E_3 \text{ OR } \ldots \text{ OR } E_m). \qquad (9.3)$$

Corollary 9.1. *Corollary of Disjoint Outcomes*
If (9.3) is true, then

$$e = (e_1 \text{ OR } e_2 \text{ OR } e_3 \text{ OR } \ldots \text{ OR } e_m). \qquad (9.4)$$

Proof. Each sub-event brings forth one outcome at a time or expresses a mutually exclusive truth. This means that e coincides with one outcome out of m potential outcomes each time.

The relation AND describes the *combined event* whose sub-events function all together to achieve the overall upshot. In the epistemic context, the propositions denoting complementary truths make the complete phrase.

Definition 9.3. *The combined event has this structure*

$$\hat{E} = (E_1 \text{ AND } E_2 \text{ AND } E_3 \text{ AND } \ldots \text{ AND } E_m). \qquad (9.5)$$

Corollary 9.2. *Corollary of Combined Outcomes*
If (9.5) holds, then

$$e = (e_1 \text{ AND } e_2 \text{ AND } e_3 \text{ AND } \ldots \text{ AND } e_m). \qquad (9.6)$$

Proof. The sub-events of (9.5) contribute to the common goal or all together are true. This means that all sub-outcomes jointly contribute to the physical outcome or global sentence e.

9.1.4 Probabilists employ *propositional algebra* and *set algebra* to detail (9.4) and (9.6). In particular, the first algebra formalizes the relationships AND and OR using the logical conjunction and disjunction; the second employs the intersection and union operations. I neglect this topic which is amply treated in the literature.

I confine myself to highlighting how the outcomes (9.4) and (9.6) descend from the structures (9.3) and (9.5), respectively, which yield the analytical descriptions of *e* in harmony with property 9.1.

9.2 Disjoint Events

While the elementary $\mathbf{E}^{(x)}$ does not say anything about the missing x, the structure in OR gives the full account of all the possible upshots.

9.2.1 The sub-events of the disjoint $\hat{\mathbf{E}}$ usually own the following property:

Definition 9.3. *When two or more events share the initial elements (i, r) and have different outcomes, they are called variants.* (9.7)

The reader can note how the elements *i* and *r* make explicit the concepts of 'initial conditions' and 'variant' (Sections 3.2 and 4.4) that the current literature takes as 'tacit axioms'

9.2.2 We call *sample space* Ω the set of *all the possible alternative* outcomes $e_1, e_2, e_3 \ldots e_N$ of the disjoint $\hat{\mathbf{E}}$. The following exposition holds:

Definition 9.4. *The disjoint structure* (9.3), *equipped with variant sub-events, is complete when* $m = N$, *and is incomplete if* $m < N$.

$$(9.8)$$

The complete disjoint structure $\hat{\mathbf{E}}$ turns out to be the most significant model to formalize aleatory phenomena and uncertain beliefs.

Theorem 9.1. *Theorem of the Complete Structure (TCS) If* (9.3) *is complete, then the global outcome e is certain*

$$e = (e_1 \text{ OR } e_2 \text{ OR } \ldots \text{ OR } e_N). \tag{9.9}$$

And the structure $\hat{\boldsymbol{E}}^{[N)}$ is 'quasi-certain'

$$\hat{\boldsymbol{E}}^{[N)} = [\mathbf{E}^{(e1)} \text{ OR } \mathbf{E}^{(e2)} \text{ OR } \ldots \text{ OR } \mathbf{E}^{(eN)}]. \qquad (9.10)$$

Proof. The sub-events are variant, hence we obtain

$$\hat{\mathbf{E}}^{[N)} = [\mathbf{E}^{(e1)} \text{ OR } \mathbf{E}^{(e2)} \text{ OR } \ldots \text{ OR } \mathbf{E}^{(eN)}] =$$
$$= [(i,r;e_1) \text{ OR } (i,r;e_2) \text{ OR } \ldots \text{ OR } (i,r;e_N)]. \quad (9.11)$$

The event $\hat{\mathbf{E}}^{[N)}$ is complete, namely it surely brings forth e, and in conformity with (8.4), we have

$$\hat{\mathbf{E}}^{[N)} = [\mathbf{E}^{(e1)} \text{ OR } \mathbf{E}^{(e2)} \text{ OR } \ldots \text{ OR } \mathbf{E}^{(eN)}] =$$
$$= [(i,r;e_1) \text{ OR } (i,r;e_2) \text{ OR } \ldots \text{ OR } (i,r;e_N)] =$$
$$= [(i,r) - (e_1 \text{ OR } e_2 \ldots \text{ OR } e_N)] =$$
$$= (i, r - e). \qquad (9.12)$$

The upper two lines of (9.12) describe the random components of $\hat{\mathbf{E}}^{[N)}$, the lower lines depict the macroscopic parts that are certain. It is evident how the overall structure is twofold: $\hat{\mathbf{E}}^{[N)}$ is certain and uncertain at the same time. The single result is random, while the overall result e is certain. Therefore, we conclude that $\hat{\mathbf{E}}^{[N)}$ is *quasi-certain* and the brackets [ ) symbolize the two features.

9.2.3 In the epistemic context, if one establishes all the possible conclusions from the premise, then the global statement is certainly true.

Example. Human lifespan includes three periods: youth, maturity, and senility (*yms*). Suppose Peter is living but you ignore his age. Three uncertain propositions give the certain conclusion

$$\hat{\mathbf{E}}^{[yms)} = (\text{"Peter is alive", Logical induction} - \text{"Peter is}$$
$$\text{young" OR "Peter is an adult" OR "Peter is old"}).$$

A self-explanatory example is sufficient to illustrate $\hat{\mathbf{E}}^{[N)}$ in the physical context.

Example. When you throw a die, one of the numbers $1, 2, \ldots, 6$ will certainly come out even if you cannot predict the result of a single throw

$$\hat{\mathbf{E}}^{[123456]} = [i, r - (1 \text{ OR } 2 \text{ OR } \ldots \text{OR } 6)] = [\mathbf{E}^{(1)} \text{ OR } \ldots \text{OR } \mathbf{E}^{(6)}].$$

9.2.4 The structure $\hat{\mathbf{E}}^{[N]}$ gives the accurate description of material and mental events, moreover $\hat{\mathbf{E}}^{[N]}$ can round out the presentation of a generic random occurrence in the following manner.

Theorem 9.2. *Theorem of the Minimal Configuration (TMC) Suppose* $\mathbf{E}^{(e)} = (i, r; e)$ *is any elementary random event, then* $\mathbf{E}^{(e)}$ *is a part of the following structure:*

$$\hat{\mathbf{E}}^{[2]} = [\mathbf{E}^{(e)} \text{ OR } \mathbf{E}^{(\bar{e})}]. \tag{9.13}$$

where $\bar{e}$ is the outcome complementary to e.

Proof. Given the aleatory upshot e, we can define whatever single outcome or group of outcomes that are alternative to e

$$\bar{e} = \text{NOT } e. \tag{9.14}$$

The outcome $\bar{e}$ belongs to $\mathbf{E}^{(\bar{e})}$, which is a variant alternative of $\mathbf{E}^{(e)}$

$$\mathbf{E}^{(\bar{e})} = (i, r; \bar{e}). \tag{9.15}$$

The ensemble $e = (e \text{ OR } \bar{e})$ is complete, and $\mathbf{E}^{(\bar{e})}$ and $\mathbf{E}^{(e)}$ make *the minimal quasi-certain structure* (9.13).

TMC makes it so that even the most mysterious and unpredictable occurrence can be placed within the 'cornice (=frame)' $\hat{\mathbf{E}}^{[2]}$ which has some traits of certainty and enlightens even the most enigmatic circumstances. The minimal configuration (9.13) demonstrates the importance and spreading of the quasi-certain model in the probability sector.

Modern mathematicians are familiar with this method; they call e and $\bar{e}$ as *success* and *failure*, respectively, inside the Bernoulli scheme.

9.2.5 The next theorem concludes the description of the disjoined events.

Theorem 9.3. *Theorem of the Incomplete Structure (TIS) If* (9.3)
is incomplete, then it is random

$$\hat{E}^{(m)} = [E^{(e1)} \text{ OR } E^{(e2)} \text{ OR } \dots \text{ OR } E^{(em)}], \quad m < N. \qquad (9.16)$$

Proof. The global result $e = (e_1 \text{ OR } e_2 \dots \text{ OR } e_m)$ is not complete,
and an experiment may output one result out of the $(N - m)$ results
that (9.16) brings forth. The link r placed between i and the output
e is not systematically established, it may fail, and $\hat{E}^{(m)}$ is uncertain
in harmony with definition 8.3

$$[E^{(e1)} \text{ OR } E^{(e2)} \text{ OR } \dots \text{ OR } E^{(em)}] =$$
$$= [(i, r; e_1) \text{ OR } (i, r; e_2) \text{ OR } \dots \text{ OR } (i, r; e_m)] =$$
$$= [i, r; (e_1 \text{ OR } e_2 \dots \text{ OR } e_m)] =$$
$$= (i, r; e) = \hat{E}^{(m)}. \qquad (9.17)$$

Example. A die has six numbers, and $\hat{E}^{(even)}$, which gets an even
number, is random since an odd number might come out

$$\hat{E}^{(even)} = [E^{(2)} \text{ OR } E^{(4)} \text{ OR } E^{(6)}].$$

As a final brief, I underscore that the theorems and corollaries are
proved on the basis of properties 7.1, 7.2, and 9.1, and not on the
basis of abstract properties or using set algebra.

9.3 Graphs

Graphs are visual tools employed to represent pairwise relationships
between objects. A graph consists of *vertices* which are connected by
edges that may or may not be *directed* (Ehrig, 1979).

9.3.1 A graph depicts the event; in detail, the vertices describe the
initial and final parts of **E** and the edge illustrates the process r
connecting them. The theory of graphs not only conforms with the
present comprehensive theory, but also depicts the intrinsic characteristics of the components. The arrow renders visible the con-
nective role of r, the dots symbolize the passive roles of i and e
(Figure 9.1).

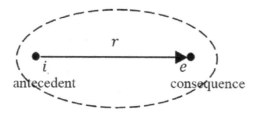

Figure 9.1. The elementary event graph.

Example. The tossed coin can be visualized in the following manner:

Graphs can even represent mental events such as the following.

Examples. The rational inference (8.1) analyzed by Keynes has the ensuing form

Carnap's scheme (8.2) can be visualized this way

The *tree diagram* depicts disjoined and combined events.

9.3.2 Graphs prove to be effective in education (Samaniego, 2014); managers draw *decision trees* (Posner, 2010), professionals use *risk trees,* etc. Graphs have also invaded the probability sector. *Bayesian networks* represent sets of variables and their conditional dependencies (further comments will follow in Chapter 15).

The probability tree diagram is a way of showing combinations of two or more elementary events that helps the analysis of events (Figure 9.2). For instance, several school exercises of probability are basically exercises of event analysis (Rocchi, 2003, Chapter 15).

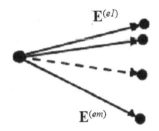

Figure 9.2. The composite event graph.

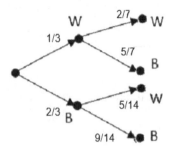

Figure 9.3. Extractions of white and black marbles.

Example. A bag contains 5 white marbles and 10 black marbles. William selects two marbles without replacement.

(a) What is the probability of selecting two white marbles?
(b) What is the probability of selecting a white and a black marble?

The anatomy of the composite event constitutes the core of this exercise (Figure 9.3). Once the first step is over with the aid of the tree graph, the calculations turn out to be trivial for the student

$$P(\mathbf{a}) = \frac{1}{3} \bullet \frac{2}{7} = \frac{2}{21}.$$

$$P(\mathbf{b}) = \frac{1}{3} \bullet \frac{5}{7} + \frac{2}{3} \bullet \frac{5}{14} = \frac{10}{21}.$$

9.3.3 Probabilists ***ordinarily draw*** graphs on empirical basis, instead ***the three parts of graphs*** are perfectly consistent with

(i, r, e). Furthermore, the present framework allows the employment of complementary visual representations such as the *Venn diagrams* and the *semantic trees* that collect propositions (Section 9.1.4).

Synopsis

a. The elementary event **E** has the property of occurring and does so by means of the specialized elements i, r, and e.

b. Events achieve three main degrees of action and cognition.

c. The compound event **Ê** occurs due to its sub-events.

9.4 Summary of the Structural Models

It seems fine to collect the main structures discussed in Chapters 8 and 9.

9.4.1 The **elementary certain** event $\mathbf{E}^{[e]}$ systematically carries out the result. Once the antecedents are settled, the conclusion follows surely

$$\mathbf{E}^{[e]} = (i, r - e). \tag{8.4}$$

9.4.2 The **elementary impossible** event $\mathbf{E}^{\{e\}}$ cannot carry about the expected result

$$\mathbf{E}^{\{e\}} = (i, r||e). \tag{8.5}$$

9.4.3 The **elementary random** event $\mathbf{E}^{(e)}$ does not furnish e systematically. Given the premises, the outcome e is unpredictable

$$\mathbf{E}^{(e)} = (i, r; e). \tag{8.6}$$

9.4.4 The components of the **disjoint** event happen one after the other or express mutually exclusive truths

$$\hat{\mathbf{E}} = (\mathbf{E}_1 \text{ OR } \mathbf{E}_2 \text{ OR } \mathbf{E}_3 \text{ OR } \ldots \text{OR } \mathbf{E}_m). \tag{9.3}$$

9.4.5 The components of the **combined** event concur to reach the common goal or all together are true

$$\hat{\mathbf{E}} = (\mathbf{E}_1 \text{ AND } \mathbf{E}_2 \text{ AND } \mathbf{E}_3 \text{ AND } \ldots \text{AND } \mathbf{E}_m). \tag{9.5}$$

9.4.6 The **complete disjoint structure** $\hat{\mathbf{E}}^{[N)}$ is quasi-certain since it certainly delivers the result e while the outcome of each trial cannot be forecast

$$\hat{\mathbf{E}}^{[N)} = [\mathbf{E}^{(e1)} \text{ OR } \mathbf{E}^{(e2)} \text{ OR } \dots \text{ OR } \mathbf{E}^{(eN)}]. \tag{9.10}$$

9.4.7 The **incomplete disjoint structure** $\hat{\mathbf{E}}^{(m)}$ is random since the outcome e is random

$$\hat{\mathbf{E}}^{(m)} = [\mathbf{E}^{(e1)} \text{ OR } \mathbf{E}^{(e2)} \text{ OR } \dots \text{ OR } \mathbf{E}^{(em)}]. \tag{9.16}$$

The triad allows us to anatomize the spectrum of circumstances located between uncertainty and certainty. The structures make us aware of the variety of objects typical of 'the logic of the uncertain' and pave the way for analyzing the multifaceted nature of probability.

References

Ehrig H. (1979). Introduction to the algebraic theory of graph grammars (a survey), In V. Claus, H. Ehrig, and G. Rozenberg (eds.), *Graph-Grammars and Their Application to Computer Science and Biology. Graph Grammars*, Vol. 73 (Springer, Berlin; New York).

Posner K.A. (2010). *Stalking the Black Swan: Research and Decision Making in a World of Extreme Volatility* (Columbia Business School, New York).

Rocchi P. (2003). *The Structural Theory of Probability; New Ideas from Computer Science on the Ancient Problem of Probability Interpretation* (Kluwer/Plenum, New York).

Samaniego F.J. (2014). *Stochastic Modeling and Mathematical Statistics: A Text for Statisticians and quantitative scientists* (CRC Press, Boca Raton, FL).

Chapter 10

Measuring the Events

The present theory is grounded on the idea that an event is something that happens (Property 7.1) in the physical world or the human mind (Section 8.1).

10.1 The Calculus of the Occurrences

This capability which defines the concept of event ranks prominently and must be accurately measured.

10.1.1 Definitions 8.1 and 8.2 illustrate events that either certainly occur or never occur, while the event 8.3 can be 'placed in the middle'. Therefore, the numeric parameter which quantifies the existence of events varies between two extreme values.

Definition 10.1. *Probability is the normalized theoretical parameter that qualifies the ability of the event to occur*

$$0 \le P(\mathbf{E}) \le 1, \quad P \in \mathbb{R}. \tag{10.1}$$

The adjective 'theoretical' means that $P(\mathbf{E})$ is obtained from calculations, and the specific characteristics of $\mathbf{E}$ yield the numerical values as specified in what follows.

10.1.2 The more likely the event *happens or will happen or potentially is able to happen,* the higher the probability. The typologies of events, commented in Chapter 8, allow us to detail this criterion.

Definition 10.2. *The value of certainty – The element r of the certain event systematically connects the input to the output and the maximum value qualifies the existence of* $\mathbf{E}^{[e]}$

$$P[\mathbf{E}^{[e]}] = P(i, r - e) = 1. \tag{10.2}$$

Definition 10.3. *The value of impossibility – There is no link between the antecedent and the consequent in the impossible event. The minimum qualifies the non-existence of* $\mathbf{E}^{\{e\}}$

$$P[\mathbf{E}^{\{e\}}] = P(i, r \,||\, e) = 0. \tag{10.3}$$

The impossible event has probability zero, but the events with probability zero are not necessarily impossible. This specification typically descends from continuous probability distributions, and is consistent with the present framework. For the sake of simplicity, we restrict attention to discrete probability spaces.

Definition 10.4. *The value of randomness – The indeterministic event does not systematically happen and an intermediate value qualifies the occasional connections of the premise with the outcome*

$$0 < \{P[\mathbf{E}^{(e)}] = P(i, r; e)\} < 1. \tag{10.4}$$

All theories of probability – despite their variety – conform to the above capstones. For instance, even imprecise and negative probabilities share the values of certainty and impossibility.

10.1.3 The event gives rise to the probability assessment and as such involves the very essence of P. Using Spinoza's criterion (Section 6.2.2), we can say that $\mathbf{E}$ turns out to be the essential reason for $P(\mathbf{E})$ and thus the election of the event as primary theoretical notion is correct in point of logic.

   This study assumes that the event has the 'propensity' to happen, no matter it is material or mental. This perspective is consistent with Popper who thinks of probability as the measure of the propensity, disposition, or tendency of a given situation.

10.2 Continuous and Discrete Probabilities

A probability distribution can assume a countable, usually finite, number of values, or otherwise can assume an infinite number of different values. This is the first and most common criterion for classifying *discrete and continuous probabilities*.

A second criterion, extremely significant in this study, is formulated as follows. Section 10.1 assigns a high value of P to certain events, a low value to impossible events, and a real number between zero and one to aleatory events. When probability qualifies three distinct cases, it becomes a *discrete variable*. In summary, we fix this classification:

Property 10.1

- *When P varies continuously within the interval $(0, 1)$, it is continuous.*
- *When P is either one, zero, or a generic decimal, it is discrete.*

$$(10.5)$$

Probability can be employed as a continuous or as a discrete variable and the second part of property 10.1 will be systematically used in exploring classical and quantum physics (Chapters 16 and 17).

10.3 Probability of the Outcome

A mathematical parameter turns out to be useful because it does not qualify only one item but several different items.

10.3.1 Speaking in general, the quantity R can gauge two elements with close relations such as the whole X and its part x, which normally do not have the same size. For instance, the cable X is made of several copper wires x, and the electrical resistances are not equal

$$x \subset X$$

$$R(X) < R(x). \tag{10.6}$$

It may be that the attributes of the whole and the part have the same size. For example, the cabin x belongs to the ship X which is sailing on the sea. Both the ship and the cabin have identical speeds

$$x \subset X \qquad \qquad (10.7a)$$

$$V(X) = V(x). \qquad \qquad (10.7b)$$

Property 7.2 holds that if the event takes place, its outcome also takes place and vice versa; therefore, the measure of the event's existence is the same as the measure of the outcome's existence. If *the event is impossible, random, or certain, so is the outcome* (Property 7.2), this logical implication also works the other way round. We can assign the probability of e to the probability of the corresponding event and vice versa.

Property 10.2

$$e \subset \mathbf{E} \qquad \qquad (10.8a)$$

$$P(e) = P(\mathbf{E}). \qquad \qquad (10.8b)$$

For the sake of convenience, one can calculate $P(e)$ in place of $P(\mathbf{E})$; otherwise, one can fix the latter and extend the value to the former.

10.3.2 Nobody confuses the ship with the cabin in consequence of equality (10.7b), instead mathematicians who have paid little attention to the event and its upshot (Chapter 4), have merged the first with the second, they have mixed together the part with the whole.

10.4 One Definition and a Variety of Nuances

The Latin adjective '*probabilis* (probable)' was born in jurisprudence; ancient lawyers, judges, and men of law were familiar with it (Chapter 2).

10.4.1 'Probabilis' had a double significance from its early beginnings, it stood for both '*verisimilis* (truth-like)' and '*credibilis* (credible)'. The first attribute refers to a clue, a witness, or evidence complying with reality; the second regards human knowledge and thinking. The physical and cognitive viewpoints (Section 8.3) emerged from the very birth of the probability conceptualization

(Deman, 1933). The first position assesses something real, the second *refers to judgments* and is about believability.

This double-dealing continued throughout the centuries and was recognized many times. Poisson made a distinction between 'probabilité' as 'individual, subjective, probability linked to a particular person' and 'chance' that is completely independent of the human will. Cournot classified *'subjective'* and *'objective'* probabilities (Daston, 1988). More recently Hacking (2006) divides the *'epistemological'* probabilities from the *'aleatoric'* ones and Gillies (1973) talks about *'logical'* and *'scientific'* probabilities.

10.4.2 The two perspectives generate a variety of shades. Scholars oriented to the ontological approach see a probable fact as

- a possibility,
- a chance,
- a potential situation,
- a verisimilitude, etc.

Those oriented to the epistemic view see a probable idea as

- something credible,
- something plausible,
- something true,
- an expectation, etc.

Tons of books have been written about these nuances which seem impossible to reconcile. Partial theories push researchers toward conflicting positions; instead, the present structural framework subsumes and embraces all the viewpoints.

10.4.3 In this theory, the meaning of $P(\mathbf{E})$ is the *consequence of* $\mathbf{E}$. The triad describes material and mental events; accordingly, probability has either ontological–physical value or epistemic–cognitive value, respectively. The significance no longer comes from philosophical convictions or personal will. It derives from the event typology; namely, $\mathbf{E}$ is a fact, a concrete possibility, a chance, etc. or otherwise a plausible credence, a belief, etc.

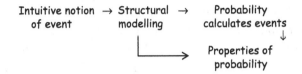

Figure 10.1. Deductive stages of probability theorization.

10.4.4 The various properties of P will be inferred from **E** (Figure 10.1), which has been stated as the primary notion, therefore, all the conclusions must be deduced from it.

In particular, the next chapter will prove the basic formulas of probability by exploiting the features of events.

Synopsis

a. The elementary event **E** has the property of occurring and does so by means of the specialized elements i, r and e.

b. Events achieve three main degrees of action and cognition.

c. The compound event **Ê** occurs due to its sub-events.

d. Probability serves as the measure of the event existence.

References

Daston L. (1988). *Classical Probability in the Enlightenment* (Princeton University Press, Princeton, NJ).

Deman T. (1933). Probabilis, *Revue des Sciences Philosophiques et Théologiques*, 22, 260–290.

Gillies D. (1973). *An Objective Theory of Probability* (Routledge, London).

Hacking I. (2006). *The Emergence of Probability: A Philosophical Study of Early Ideas about Probability, Induction and Statistical Inference*, 2nd edition (Cambridge University Press, Cambridge).

Chapter 11

Formulas to be Proven

David Hilbert (1902) presented a list of the problems that were unsolved in the early 1900s. The sixth problem was to formulate rigorously those branches of physics in which mathematics is prevalent, and probability theory ranked first because it lacked a coherent body of proofs.

This chapter will prove formulas (2.1), (2.2), and (2.3) which bettors and gamblers contrived, on the basis of properties 7.1, 7.2, and 9.1, and the subsequent achievements. This book follows the phenomenological logic, it analyzes the modes of being of events and how they unfold. This logic ignores the usual techniques of demonstration which focus on the outcomes and formalize them as subsets or propositions.

11.1 Equiprobability

Geymonat and Costantini (1982) denote Laplace's theory as '*dogmatic*', because it is characterized by an unproven assertion. The very formula (2.1), a cornerstone of probability calculus, is still waiting for a rigorous proof.

Theorem 11.1. *Classical Theorem (CT) Suppose the sub-events of* $\hat{\mathbf{E}}^{(m)}$ *are equally likely*

$$\hat{\mathbf{E}}^{(m)} = [\mathbf{E}^{(e1)} \text{ OR } \mathbf{E}^{(e2)} \text{ OR } \dots \text{ OR } \mathbf{E}^{(em)}]. \qquad (11.1)$$

Then the number m $(m < N)$ of favorable cases divided by the number N of all the possible cases gives probability

$$P[\hat{\mathbf{E}}^{(\mathbf{m})}] = \frac{m}{N}. \tag{11.2}$$

Proof. Each sub-event contributes to the occurrence of the compound event (Property 9.1). The generic $\mathbf{E}^{(ej)}(j = 1, 2 \ldots m)$ gives the possibility for $\hat{\mathbf{E}}^{(\mathrm{m})}$ to come into being, hereupon the higher m and the more likely $\hat{\mathbf{E}}^{(\mathrm{m})}$ occurs. The sub-events are equiprobable and offer identical support for $P[\hat{\mathbf{E}}^{(\mathrm{m})}]$, so $P[\hat{\mathbf{E}}^{(\mathrm{m})}]$ is proportional to the number m of sub-events up to a constant

$$P[\hat{\mathbf{E}}^{(\mathbf{m})}] \bullet c = m, \quad c > 0. \tag{11.3}$$

From (11.3), we obtain

$$P[\hat{\mathbf{E}}^{(\mathbf{m})}] = m/c, \quad c > 0. \tag{11.4}$$

In the case m equals N, $\hat{\mathbf{E}}^{(N)}$ surely happens due to the theorem of the complete structure

$$1 = \frac{N}{c}, \quad c > 0.$$

Thus, c equals the grand total N usually called *number of possible cases*

$$c = N. \tag{11.5}$$

Put (11.5) into (11.4), and (11.2) is proved.

11.2 Addition and Multiplication Rules

Theorists often use the additivity property of frequency to justify the additivity axiom, next they use this axiom to prove the multiplication rule.

Theorem 11.2. *Theorem of Addition (TA)* **S***uppose* $\hat{\mathbf{E}}^{(\mathrm{m})}$ *is an incomplete structure whose sub-events are variant and mutually exclusive*

$$\hat{\mathbf{E}}^{(\mathbf{m})} = [\mathbf{E}^{(e1)} \,\mathrm{OR}\, \mathbf{E}^{(e2)} \,\mathrm{OR}\, \ldots \,\mathrm{OR}\, \mathbf{E}^{(em)}]. \tag{9.16}$$

Then the probability function which qualifies this structure is the sum

$$P[\hat{\mathbf{E}}^{(m)}] = P[\mathbf{E}^{(e1)}] + P[\mathbf{E}^{(e2)}] + P[\mathbf{E}^{(e3)}] + \cdots + P[\mathbf{E}^{(em)}]. \quad (11.6)$$

Proof. Property 9.1 maintains that $\hat{\mathbf{E}}^{(m)}$ occurs due to each sub-event, thus it is logical to conclude that the greater $P[\mathbf{E}^{(ej)}]$ $(j = 1, 2, \ldots m)$ is, the greater is $P[\hat{\mathbf{E}}^{(m)}]$.

As second, the sub-events are alternative, hence the higher the number m, the higher is the possibility of $\hat{\mathbf{E}}^{(m)}$ to come about, because each sub-event provides independent support. This pair of remarks can be specified as follows:

1. $P[\hat{\mathbf{E}}^{(m)}]$ *is an increasing monotonic function of* $P\ [\mathbf{E}^{(ej)}]$ *with* $j = 1, 2, \ldots m$.
2. $P[\hat{\mathbf{E}}^{(m)}]$ *is an increasing monotonic function of* m.

By adding positive natural numbers, the result increases if the generic addend becomes larger and the number of addends increases, thus (11.6) is proved.

Corollary 11.1. *Corollary of the Sample Space*
If $m = N$, *the quasi-certain event verifies this summation*

$$P[\hat{\mathbf{E}}^{(N)}] = 1.$$

Proof. From (11.6), we get

$$P[\hat{\mathbf{E}}^{(N)}] = \sum_{k}^{N} P\left[\mathbf{E}^{(ek)}\right] = 1.$$

The theorem of the complete structure (Section 9.2) proves that $\hat{\mathbf{E}}^{(N)}$ certainly provides e, that is, $P[\hat{\mathbf{E}}^{(N)}]$ is one. TCS and corollary 11.1 develop different reasonings, yet they obtain the same result.

Theorem 11.3. *Theorem of Multiplication (TM)* ***S****uppose* $\hat{\mathbf{E}}^{(m)}$ *is a combined and incomplete event whose sub-events are variant and* *independent*

$$\hat{\mathbf{E}}^{(m)} = [\mathbf{E}^{(e1)} \text{ AND } \mathbf{E}^{(e2)} \text{ AND } \ldots \text{ AND } \mathbf{E}^{(em)}]. \qquad (9.5)$$

Then the probability function which qualifies this structure is the multiplication

$$P[\hat{\mathbf{E}}^{(m)}] = P[\mathbf{E}^{(e1)}] \bullet P[\mathbf{E}^{(e2)}] \bullet P[\mathbf{E}^{(e3)}] \bullet \ldots \ldots \bullet P[\mathbf{E}^{(em)}]. \quad (11.7)$$

Proof. Property 9.1 asserts that $\hat{\mathbf{E}}^{(m)}$ occurs by means of each sub-event, thus the greater $P[\mathbf{E}^{(ej)}]$ $(j = 1, 2, \ldots m)$ is, the greater is $P[\hat{\mathbf{E}}^{(m)}]$.

Secondly, it only takes one sub-event for the entire event to fail; therefore, the capability of coming about for $\hat{\mathbf{E}}^{(m)}$ declines when m becomes greater. The two remarks can be specified as follows:

1. $P[\hat{\mathbf{E}}^{(m)}]$ *is an increasing monotonic function of* $P[\mathbf{E}^{(ej)}]$ *with* $j = 1, 2, \ldots m$.
2. $P[\hat{\mathbf{E}}^{(m)}]$ *is a decreasing monotonic function of* m.

The multiplication operation of positive natural numbers obtains a result which increases as the multipliers become greater and decreases as the number m grows because the multipliers are less than one; (11.7) is proven.

As a final brief, it is worth pointing out that properties 7.1, 7.2, and 9.1 make the backbone of the demonstrations just illustrated; they also are in harmony with the purposes of this construction that aims to infer the properties of P from the phenomenological perspective.

11.3 Proving Postulated Truths

The results obtained in this book aid us to justify the axioms shared by probability theories as follows:

- Definition 10.1 legitimizes *the non-negativity axiom* (point III, Section 3.2.2).

- The corollary of the sample space spells out *the normalization axiom* (point IV, Section 3.2.2).
- The theorem of addition makes explicit *the summation axiom* (point V, Section 3.2.2).

The current structural theory is consistent with modern theories in mathematical terms and confirms the metaphor 'Newton–Lagrange' put forward in Section 6.3.

References

Geymonat L. and Costantini D. (1982). *Filosofia della Probabilità* (Feltrinelli Editore, Milano).

Hilbert D. (1902). Mathematical problems, *Bulletin of the American Mathematical Society*, 8(10), 437–479.

Chapter 12

How to Test Probability

Testing is a well-established form of control which is typical of the scientific method and consists of a series of experiments done under exact conditions that also rely on feedback procedures (Figure 12.1).

A theory is not necessarily right and Popper (1963) emphasizes the importance of controlling any scientific claim. A truly scientific theory must offer the possibility of being falsified on *the logical plane and the experimental plane* alike. Predictions must be in agreement with observations and for Popper an empirical outcome opposed to the forecasts is sufficient to confute the entire theoretical construction. He concludes that the essential difference between science and pseudoscience is that the second seeks confirmations and the first seeks experimental falsifications.

The need for testing is central for the frequentists, while other schools of thought overlook this need. The present theory openly declares following the phenomenological approach (Section 6.2), therefore empirical data must corroborate or disprove the theoretical parameter P. The definitions and theorems illustrated in the previous pages must find a precise correspondence with facts in accordance with the scientific method.

12.1 Experimental Control

When one means to check the theoretical variable K, the first duty is to establish the empirical parameter k which corresponds to K.

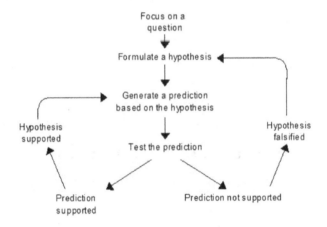

Figure 12.1. Feedback procedure for testing.

Only if k has features symmetrical to K, can the control of K be considered *valid*.

12.1.1 Chapter 10 does not leave any shadow of doubt about the significance of probability; $P(\mathbf{E})$ qualifies the existence of $\mathbf{E}$ in the abstract and the frequency $F(\mathbf{E})$ does the same in practice

$$F[\mathbf{E}(i,r,e)] = \frac{q}{n}, \quad n \geq 1, \tag{12.1}$$

where q is the number of occurrences of $\mathbf{E}$ in n experiments. So, we conclude the following.

Validity Criterion (VC) *The relative frequency is the experimental counterpart of probability*

$$P[\mathbf{E}(i,r,e)] \leftrightarrow F[\mathbf{E}(i,r,e)]. \tag{12.2}$$

The literature recognizes this conceptual alignment and F is frequently named as '*empirical probability*'.

12.1.2 Speaking in general, the experimental control may meet difficulties in science; it may run into limitations or even prove to be impossible to execute. Experimentalists encounter material impediments in astronomy, ethical constraints in psychiatry, personal risks in vulcanology, inexecutable operations in economics, and so on. It is necessary to examine carefully how and when experts can use criterion (12.2).

12.2 Testing Random Events

Deterministic phenomena are stable and sometimes a single test is sufficient to disprove a deterministic event. This is just to conclude that criterion (12.2) does not raise great difficulties in this area, at least in principle.

Fluctuating empirical data proves to be much harder to control.

12.2.1 Let us consider not the space Ω of possible events, but the space Θ of events that factually occur in a certain period of time or context. In the space Θ we define the event $\mathbf{E}_n^{(e)}$ tested n times, in fact testing does not regard $P[\mathbf{E}^{(e)}]$, which has generic meaning in the physical world, but $P[\mathbf{E}_n^{(e)}]$ which specifies the experimental conditions.

Popper focuses on the aleatory events that is tested once and infinitely many times. The long-term event $\mathbf{E}_\infty^{(e)}$, close to the *collective* by von Mises and the *series* by Venn, turns out to be a very natural reference for experimental sciences exploring the general laws of Nature (de Groot, 1975). Decision-making, financial forecasts, risks assessment, and other applications usually involve the single contingent fact $\mathbf{E}_1^{(e)}$. The relationship between the two turns out to be somewhat evident.

Definition 12.1. *The long-term event consists of elementary events, each of which is tested once*

$$\mathbf{E}_\infty^{(e)} = ((\mathbf{E}_{11}^{(e)}, \mathbf{E}_{12}^{(e)}, \mathbf{E}_{13}^{(e)}, \ldots)), \tag{12.3}$$

where $\mathbf{E}_\infty^{(e)}$ and $\mathbf{E}_{1h}^{(e)} (h = 1, 2, 3 \ldots)$ are variant. The double brackets indicate that the sub-events make a random series emitting *independent and identically distributed outcomes*.

The first theorem deals with the test of $\mathbf{E}_\infty^{(e)}$.

Theorem 12.1. *Theorem of Large Numbers (TLN)*
Consider the long-term event (12.3), the relative frequency approaches the probability when the number of trials tends to infinity

$$F[\mathbf{E}_n^{(e)}] \xrightarrow{a.s.} P[\mathbf{E}_\infty^{(e)}], \quad as \ n \to \infty. \tag{12.4}$$

Proof. Available in (Rocchi, 2014).

Example. Suppose a player flips a coin 10,000 times and gets heads 5,231 times. The relative frequency $F[\text{Win}_{10000}^{(heads)}] = 0.52$ approaches

$$P[\text{Win}_{10000}^{(heads)}] = 0.50.$$

12.2.2 This theorem ensures that an experimental system can corroborate or invalidate the probability of a long-term event, at least in principle.

The ideal configuration (12.3) requires an infinite number of trials, which is unachievable; however, the increasing precision ensured by the statistical convergence (12.4) enables the operator to adopt an approximation criterion. The higher the n, the higher the accuracy of the probability control.

The theorem of large numbers proves that $P[\mathbf{E}_\infty^{(e)}]$ can be tested; therefore, it is a *physical and objective quantity*, independent of the human mind. The *frequentist probability* $P[\mathbf{E}_\infty^{(e)}]$ exhibits the following properties which are mutually related:

$$\mathbf{E}_\infty^{(e)} \text{ is a } material \text{ event and}$$

experts assess $P[\mathbf{E}_\infty^{(e)}]$ from the *ontological viewpoint*.

$P[\mathbf{E}_\infty^{(e)}]$ is a *testable quantity* and has *objective meaning*.

$$(12.5)$$

12.2.3 The theorem of large numbers calls the *law of large numbers* to mind, in particular the version devised by Émile Borel where the relative frequency tends toward the probability as the number of experiments grows

$$\frac{q}{n} \xrightarrow{a.s.} P, \quad as \ n \to \infty. \tag{12.6}$$

Despite a certain formal similarity, LLN and TLN pursue diverse scopes and have different contents. It may be said that they tackle the same phenomenon from distinct stances.

The law focuses on *the statistical convergence of empirical data toward the expected value* (Révész, 1967) and formulates a general mathematical property. The theorem focuses on *the experimental validation of probability* and ensures that it can be checked when the event repeats. TLN refers to the probability of $\mathbf{E}_\infty^{(e)}$ that is a very special occurrence; LLN refers to the concept of probability in the abstract, and for this reason, the conclusions of TLN differ from the conclusions of LLN.

The theorem of large numbers is alien to the epistemic environment; instead, epistemic theories share the law of large numbers because of its general mathematical significance. For instance, Keynes renames the LLN as '*stability of statistical frequencies*' and judges it of great importance for statistical induction. De Finetti (1933) proves a strong version of LLN for exchangeable variables.

12.2.4 Let us scrutinize carefully the second theorem that regulates probability testing.

Theorem 12.2. *Theorem of a Single Number (TSN) or Lower Bound Theorem.*

In a single trial, the relative frequency of a random event does not equal its probability

$$F[\mathbf{E}_1^{(e)}] \neq P[\mathbf{E}_1^{(e)}]. \tag{12.7}$$

Proof. Available in (Rocchi, 2014).

Example. Suppose a flipped coin lands on heads. We obtain $F[\text{Win}_1^{(heads)}] = 1$, which altogether unfits with the numerical value of probability

$$P[\text{Win}_1^{(heads)}] = 0.5.$$

Even if the coin lands on tails, the frequency $F[\text{Win}_1^{(heads)}] = 0$ mismatches with probability.

12.2.5 TSN demonstrates that an operator cannot substantiate the probability of the individual aleatory event. This limitation has nothing to do with inadequacies in the measuring instruments, the techniques, the procedures, or any other material impediment. TSN points out the inner nature of the probability of the single occurrence that is out of control. An expert can estimate $P[\mathbf{E}_1^{(e)}]$, but the number never conforms to experience. This defect is not a novelty in the literature; for instance, Giere (1973) repeatedly calls the attention of Popper to the "problem of the single case" and underscores its intrinsic missing correspondence with the reality.

The whole of science is exclusively compatible with everything which matches with facts and testing regulates the acceptance/rejection of any theoretical proposal whatever. If the theoretical parameter X does not comply with any experiment or is completely unverifiable, then scientists conclude X has no actual substance; thereby, it is necessary to conclude that $P[\mathbf{E}_1^{(e)}]$ is not real. This deduction perfectly matches with that of de Finetti who wrote in capital letters

$$\text{"Probability does not exist."} \tag{12.8}$$

The impossibility of falsification deprives $P[\mathbf{E}_1^{(e)}]$ of any scientific position, and TSN leads to the following precise resolution:

$$P[\mathbf{E}_1^{(e)}] \text{ must be rejected.} \tag{12.9}$$

Some commentators object that the aphorism (12.8) sounds like a *radical* statement (Galavotti, 2009), others judge it as inappropriate for the mathematical context, and so on. Personal opinions do not have space here; the theorem of a single number supports the sentences (12.8) and (12.9), thence either one disproves (12.7) or otherwise these conclusions are correct.

12.3 An Intelligent Maneuver

Many people are very interested in single facts and want to assess them by means of probability. Individual events involve the lives and professional activities of people. This deliberate focus, which emerged since time immemorial, requires circumventing the previous limitation:

How can theorists combine human desires with conclusive rejection (12.9)?

The subjectivists accepted the challenge, whereas the frequentists definitively refused to confront the question. Let us look into the conceptual '*escamotage*' devised by the former to meet the needs of people.

12.3.1 *Semiotics* teaches us that *words, numbers, pictures, films, texts, and so forth are signs,* namely, they have the capability of representing something and we ordinarily say they have *semantic properties.*

The number $P[\mathbf{E}_\infty^{(e)}]$ stands for a physical quantity. It represents something real, whereas $P[\mathbf{E}_1^{(e)}]$ does not, yet it retains semantic capabilities since it is a number, and conveys significance despite remarks (12.8) and (12.9). Because of this undeniable semantic aptitude, the subjectivists and the Bayesians reuse $P[\mathbf{E}_1^{(e)}]$, which otherwise should be discarded, by ascribing a special meaning to it (Jackson, 2020), more precisely

$P[\mathbf{E}_1^{(e)}]$ *qualifies the personal belief in the occurrence of* $\mathbf{E}_1^{(e)}$.

$$(12.10)$$

This probability, usually called *subjective* or *Bayesian*, expresses the degree of personal expectation from the single trial.

12.3.2 Subjectivists and Bayesians never mention *semiotics*, nor do they show any concern for this field of study:

How could they have exploited the semantic properties then?

Semiotics teach us that a sign has two components, called the *signifier* and the *signified*; they are best known as *form* (or *shape*) and *content* (also *meaning* or *significance*) (Appendix B)

$$\text{Sign} = \text{Signifier} + \text{Signified}.$$

The first element is the physical base of the sign; the second, is the entity, material or abstract, denoted by the signifier; in other words, the *form stands for the content* or *symbolizes the content.*

Example. Suppose you call "Jim" aloud. The sound of your voice is the material base of the signal; your friend Jim is the meaning of the sound wave.

Example. The following number is a sign:

$$67$$

The ink film placed on this sheet of paper is the signifier. The intended abstract numerical value, which is a mental entity, constitutes the meaning.

The two fundamental concepts of semiotics – form and content – prove to be very intuitive and all of us learned to use them starting in high school. The masters of subjectivism and the Bayesians exploited the semantic properties of numbers without specific studies on par with ordinary people.

Even experts from various fields share the basic semiotic tenets and are able to create interesting solutions.

Example. Mathematicians have invented various techniques to codify numbers. The number on the left side is written with the decimal base, and the number on the right is written with the binary base

$$23 \qquad 10111$$

The two signifiers symbolize the *same abstract content*, which is the mathematical numerical value. The following number has *material* content as it stands for a weight:

$$23\,\mathrm{kg}$$

Example. Hardware and software scientists have devised a large assortment of signifiers such as Magnetic Codes for hard disks, Red-Blue-Green code for display screens, the Hexadecimal code, 1D barcodes (Figure 12.2), 2D barcodes, and many others.

Mathematicians and computer engineers have exploited the concepts of shape and content without appropriate preliminary education. They are familiar with the semiotic tenets without taking special lessons, since these tenets prove to be self-explanatory.

12.4 Completing the Maneuver

The subjective–Bayesian scheme outlined by (12.10) turns out to be somewhat complex and requires further insights.

12.4.1 The number $P(\mathbf{E_1})$ qualifies the personal credence in $\mathbf{E_1}$; it does not measure a fact but the opinion of an individual about that fact. The subjectivists and the Bayesians assess the following uncertain event (Section 8.1):

$$\mathbf{E}_1^{(pr)} = (\text{Prior information, Belief; Proposition}). \qquad (12.11)$$

And do not make a category mistake because they systematically move from the physical to the psychic domain (Section 8.3)

$$P[\mathbf{E}_1^{(e)}] \Rightarrow P[\mathbf{E}_1^{(pr)}]. \qquad (12.12)$$

We place the objective and subjective schemes side by side to stress the peculiar aspects of each one. The material $\mathbf{E}^{(e)}$ has the following components:

$$i = \textit{Factual preliminary}, r = \textit{Process}, e = \textit{Output of the process}.$$
$$(12.13)$$

The mental $\mathbf{E}^{(pr)}$ is equipped with the ensuing cognitive elements:

$$i = \text{Prior information (partial)},$$

$$r = \text{Belief},$$

$$e = \text{Proposition about the belief}. \qquad (12.14a)$$

The involved person quantifies his degree of belief about 'Proposition' (pr) on the basis of 'Prior information'. The Markovian chain i-r-e shows how the first component influences the personal credence r about e. The stronger the individual's conviction about pr, the

Figure 12.2. One Dimension (1D) barcode.

stronger the link between the initial and final components of (12.11) and the more solid the existence of $\mathbf{F}_1^{(pr)}$ in the individual's mind.

12.4.2 The individual who has the perfect cognition i makes firm conclusions, namely $\mathbf{E}_1^{[pr]}$ certainly occurs or otherwise the impossible $\mathbf{E}_1^{\{pr\}}$ does not take place in the mind of the involved person. The determinate triad is symmetrical to the indeterminate triad (12.14a) except for the initial term

$$i = \text{Prior information (complete)},$$

$$r = \text{Belief},$$

$$e = \text{Proposition about the belief.} \qquad (12.14\text{b})$$

In short, when the individual begins with complete knowledge, he arrives at the precise conclusion e that is certain or otherwise impossible. When the individual has incomplete cognition, he is uncertain, and probability is a decimal.

Two individuals can take advantage of different sources of information, and even one person can enjoy a variety of sources. The decisive influence of 'Prior information' shows that some pseudo-enigmas simply do not have ground, such as:

What is the probability that the sun will rise tomorrow?

The event is single, probability is subjective, and very different answers arise depending on the available cognition and the assumptions made.

12.4.3 A subjectivist can exploit any method of calculus, since there is no particular reason why his personal credence cannot align with the modern techniques of calculus. For example, one believes that a flipped coin has a 50% chance of landing on heads even under the subjective interpretation.

In consequence of this flexibility, some are inclined to minimize the differences between the ontological and epistemic logics; instead, the present theory comprehends all the viewpoints and firmly sets an approach apart from the other. Further examples are useful to point out misunderstandings.

Example. Suppose the football team A will play against team B next Sunday. The game can be formalized as follows:

$$\mathbf{E}_1^{(A)} = (A \ and \ B \ begin \ the \ football \ game, \ A \ and \ B \ play;$$

$$A \ will \ beat \ B). \tag{12.15}$$

Peter and Paul mean to forecast whether A will beat B, but the event is unique and the probability of (12.15) 'does not exist'. They can assess only their personal views of the football game. Peter knows two strong players of team A and his thought can be formalized as follows:

$$\mathbf{E}_1^{(APeter)} = (\text{"Two players are strong"}, \ Belief; \ \text{"A will beat B"}).$$

$$\tag{12.16}$$

The credence (12.16) is influenced by Peter's familiarity with the team A and he concludes

$$P[\mathbf{E}_1^{(APeter)}] = 0.8$$

Paul has followed the entire football season where team A won 3 games out of 13, and concludes

$$\mathbf{E}_1^{(APaul)} = (\text{"A won 3 games"}, \ Belief; \ \text{"A will beat B"}).$$

$$P[\mathbf{E}_1^{(APaul)}] = 3/13 = 0.23. \tag{12.17}$$

Paul uses the *classical formula* and seems to supply an 'objective' value, but (12.17) cannot be tested, and necessarily expresses a personal appraisal. In substance, the numbers 0.8 and 0.23 quantify how solid the convictions of Peter and Paul are, which derive from their experience and cognitions.

12.4.4 Some probabilists tend to smooth the differences between frequentism and subjectivism. They tend to 'cool' controversy and divisions using a simplified jargon. For example, they claim that $P[\mathbf{E}_1^{(APeter)}]$ qualifies "the victory of A on the basis of incomplete information", instead of specifying that $P[\mathbf{E}_1^{(APeter)}]$ qualifies "Peter's judgement of the victory of A on the basis of incomplete information". The two expressions describe arguments far different in nature and placed at a distance by no means negligible.

Table 12.1. Usability of the probabilistic approaches.

Event	Material	Mental
Ontological Approach	*Applicable*	*N.A.*
Epistemic Approach	*Applicable*	*Applicable*

Some school teachers show the most similar sides of the classical and the Bayesian statistics and invite the students just to 'change the cap' when they leave objective probabilities and deal with the subjective ones. In contrast, the triad makes explicit the different perspectives of frequentists and subjectivists, and the intellectual divide between them.

12.4.5 Material facts are the typical objects of the ontological approach, while any question whatever can be the subject of epistemic inquiries (Table 12.1). The subjectivists can go beyond a specific context; they can exceed the 'hic and nunc (here and now)' context. They can assess a statement which has logical and general significance. They evaluate propositions describing thoughts, reasoning, expectations, hypotheses, and facts that may be real or even fictional. For example, a subjectivist can estimate the probability of meeting with the inhabitants of Venus or with a fairy, or the arrival of Mickey Mouse; since a person can imagine and describe whatever situation using *pr*.

The ostensible flexibility of the epistemic approach raises doubts of the following kind:

Should we conclude that the subjectivists and the Bayesians certify the authentic probability?

Subjective probability is a clever ploy to recycle $P[\mathbf{E}_1^{(e)}]$ that should be discarded; nonetheless, it remains out of control. This limitation breaks down the falsifiability criterion which, instead, is the standard that recognizes the genuinely scientific theory. It towers as a cornerstone in the scientific realm and in real life.

12.4.6 In summary, TSN proves that *the probability of a single event cannot be tested and should be discarded.* Subjectivists and Bayesians reinterpret it as the measure of *subjective credence*. They extend the epistemic approach to whatever *pr* judged on the basis of prior cognition, data, experience, etc.

$$\mathbf{E}_1^{(pr)} \text{ is a } mental \text{ event and}$$

experts assess $P[\mathbf{E}_1^{(pr)}]$ from the *epistemic viewpoint*.

$P[\mathbf{E}_1^{(pr)}]$ is *untestable* and has a *subjective meaning*. (12.18)

The reader can note how each feature (12.18) is contrary and opposed to the corresponding feature of frequentism

$$\mathbf{E}_\infty^{(e)} \text{ is a } material \text{ event and}$$

experts assess $P[\mathbf{E}_\infty^{(e)}]$ from the *ontological viewpoint*.

$P[\mathbf{E}_\infty^{(e)}]$ is *a testable quantity* and has *objective meaning*.

(12.5)

12.4.7 The present framework defines probability as 'the normalized theoretical parameter that qualifies the ability of the event to occur'. Expression (10.1) has been formulated at *the abstract level* and conforms to the mathematical schools which address probability apart from any real environment (points 1–7 in Section 2.2) (Maximov, 2014).

This chapter analyzes the relations of probability with experimental evidence, in particular it has emphasized two environments at *the practical level*: The frequentist and subjective/Bayesian environments. Consequently, probability can be calculated in the abstract, and alternatively in the frequentist or the subjective/Bayesian scheme.

Example. What is the probability of getting 3 from a rolled die? The classical formula furnishes the result $P[\mathbf{E}^{(3)}] = 1/6$, which does not refer to any specific context and is abstract. When a mathematician locates the event in the world, he has to specify whether the trial is unique $\mathbf{E}_1^{(3)}$ or repeated several times $\mathbf{E}_\infty^{(3)}$.

The present theory comprises the ontological and epistemic approaches, the frequentist and the Bayesian probabilities, the abstract and practical viewpoints, objective and subjective values, etc. At the same time, the present theory points out the assumptions and constraints which generate each variety of probability.

12.5 Further Notes About Subjectivism

The subjectivists and the Bayesians recover the probability of the single event that should be rejected; this intelligent maneuver has further implications.

12.5.1 De Finetti feels the need to formulate consistent relationships between the degrees of belief and experimental observations.

In the first step, he focuses on the concept of *exchangeability*, which has central position (de Finetti, 1930). He states that the sequence of random variables is exchangeable if the joint distribution of the sequence is unchanged by any permutation of the indices (Bernardo, 1996). Formally, the variables $x_1, x_2, \ldots, x_n$ are exchangeable if for all permutations π defined on the variables and for every finite subset of them the following equality is valid:

$$P\left(x_1, x_2, \ldots, x_n\right) = P\left(x_{\pi(1)}, x_{\pi(2)}, \ldots, x_{\pi(n)}\right). \tag{12.19}$$

The notion of exchangeability involves a judgement of complete symmetry among all the observables under consideration. The individual's beliefs about the variables are exchangeable if their order does not influence the assignment of probabilities to them.

In the second step, de Finetti proves the *representation theorem* stating that exchangeable observations are conditionally independent relative to some latent variable; in this way he relates past observations with future predictions. The representation theorem proves that if (12.19) is true, then any finite subset of variables is a random sample of some model $P(x|\theta)$ and also there is prior probability distribution $P(\theta)$ such that

$$P(x_1, x_2, \ldots, x_n) = \int_\Theta \prod_{i=1}^n P(x_i|\theta) P(\theta) d\theta. \tag{12.20}$$

where the parameter θ belongs to the prior distribution Θ, which is the limit of some function of the x_i's as $n \to \infty$. The theorem entails that the probability distribution of any infinite exchangeable sequence of Bernoulli aleatory variables is a 'mixture' of the probability distributions of independent and identically distributed sequences of Bernoulli variables.

The representation theorem is normally used in the context of Bayesian inference and plays a key role in the 'justification' of the prior distribution of the parameter of interest. It tackles the problem of relating past observations with future predictions and states that the use of relative frequencies for prediction during the induction process makes sense only in the presence of exchangeability. It is evident how the representation theorem states there is no true parameter subject to testing, there are only data and judgements in conformity with the subjective scheme.

12.5.2 All the epistemic probabilities – i.e., inferential, logical, Bayesian, subjective, etc. – presume prior information (Section 8.1) that is not so easy to define and calculate. Keynes provides insights into these difficulties (Carabelli, 1988). He accepts that some probabilities are measurable, for example, using the classical formula of Laplace, but pinpoints that several probabilities are not measurable and even cannot be compared. Chapter 12 of (Keynes, 1921) formulates a set of axioms, but immediately the author adds that epistemic probabilities are not restricted to numbers. The English mathematician sees numerical probabilities as special cases, whereas probability is intrinsically not definable, and often it cannot be quantifiable or even comparable without distorting its significance. Keynes discusses the case of the man who assesses the probability Pa given assumption a and Pb given b by means of a certain criterion. If Pa and Pb regard the same situation, Keynes notes that an expert might not be able to establish the first is greater than the second or the second is greater than the first or both are equal. The probabilities could be incomparable in magnitude because of the different intellectual perspectives a and b, which (12.11) formalizes as 'prior information'.

Keynes emphasizes how probability may not be assessed accurately or even may be intuited.

Carnap goes back to the argument and circumvents some difficulties accepting the *frequentist probability* p_2; anyway, the *logical probability* p_1 remains partially definable and he formalizes the *credibility function* to address this problem.

12.5.3 Partial 'Prior information' might vary considerably from person to person and the appraisal of uncertain events turns out to be a critical topic. The number $P[\mathbf{E}_1^{(pr)}]$ can express judgments that are subject to individual preference, and suspicions of the arbitrariness of the subjective probability have been raised early on since its inception. The co-inventors – Ramsey and de Finetti – are perfectly aware of these objections and insist greatly on the bettors' criterion to fix the correct probability value. Ramsey (1931) argues

> "The old-established way of measuring a person's belief is to propose a bet, and see what are the lowest odds which he will accept. This method I regard as fundamentally sound."

De Finetti (1964) proposes the *Dutch book* criterion to ensure the consistency of the probability assessment and adds

> "Let us suppose that an individual is obliged to evaluate the rate p at which he would be ready to exchange the possession of an arbitrary sum S (positive or negative) dependent on the occurrence of a given event E, for the possession of the sum pS; we will say by definition that this number p is the measure of the degree of probability attributed by the individual considered to the event E, or, more simply, that p is the probability of E (according to the individual considered; this specification can be implicit if there is no ambiguity)."

The Dutch book method might contain a certain degree of personal bias; nonetheless, it should be seen as a significant effort to minimize the risk of arbitrariness in probability assessment.

The Bayesian statistics provides further answers to this issue. From a distributional model of some form $P(x_i|\theta)$ and a prior $P(\theta)$, it obtains the posterior $P(\theta|x)$ at the subsequent stage. It may be said that the Bayesians – in harmony with the *representation theorem* of de Finetti – have constructed a rational guideline for improving the evaluation of probability through successive stages.

It is worth remembering that the concept of betting, which still raises doubts in some scientific sectors, dates back to the very founder of the modern probability calculus who addressed the problem of God's existence. Pascal started with the idea that reason does not lead to any precise conclusion and cannot make the individual choose either to believe or not (Morris, 1986). For Pascal, an individual has to take a probabilistic approach, he must make a wager and apply the most rational betting criterion to answer the dilemma.

12.5.4 Epistemic probabilities are substantially affected by an individual's awareness. If 'Prior information' changes, then the mental event (12.11) changes and in turn, probability. Great efforts have been made to relate the degrees of personal credence to rational and quantitative criteria. In fact, people may differ somewhat in their personal tendencies, even if they all have the same awareness. Kyburg (1961) suggests adopting an acceptance rule that establishes a fixed threshold for highly likely occurrences, but he also notices how this rule can lead to paradoxical conclusions.

Example. A player confronts the problem of a fair lottery with 1 million tickets in such a way that for each ticket x the chance of not-winning is very high: $P(x) > 0.999999$. The analysis of the lottery yields two contradictory conclusions: the player must believe that some tickets will win the lottery and at the same time all the tickets x are not winning tickets due to the high value of $P(x)$.

The elicitation of the prior becomes more complicated in the case of decision-making with multiple agents. It is necessary to select a suitable family of priors in place of assessing a single prior (Bulling, 2014). Another criticism comes from Walley (1987) who discusses the difficulty arising when conventional Bayesian analysis is presented with weak data sources. Finally, the Dempster–Shafer theory of evidence (DST) is worth mentioning. It has its origin in the work of Dempster (1967) on the use of probabilities with upper and lower bounds. DST uses *belief functions* (instead of *probability functions* as usually advocated by the Bayesians) and covers several formalisms such as the *lower probabilities* model, *Dempster's* model, the *hint* model, the *probability of modal propositions* model, and the *transferable belief* model (Beynon et al., 2000).

12.6 Impossible Testing: The Upper Bound

The theorem of a single number sets off how P mismatches with experience and the individual case constitutes the minimal configuration of probability unrealism. The reader might ask what the upper limit for unpracticable tests is.

12.6.1 The following theorem provides the maximal number of events whose probability is out of control.

Theorem 12.3. *Theorem of Upper-Bound Limit (TUBL) Let n and z be positive integers such that*

$$1 < n < z. \tag{12.21}$$

When the probability of $\mathbf{E}_n^{(e)}$ satisfies

$$P[\mathbf{E}_n^{(e)}] = 1/z. \tag{12.22}$$

Then the relative frequency of the event in n trials does not equal probability

$$F[\mathbf{E}_n^{(e)}] \neq P[\mathbf{E}_n^{(e)}]. \tag{12.23}$$

Proof. *Available in* (Rocchi, 2017).

TUBL demonstrates that empirical tests cannot control $P[\mathbf{E}_n^{(e)}]$ under the constraints (12.21) and (12.22).

Example. The probability of picking up a queen from a card deck is

$$P[\mathbf{E}^{(Q)}] = 4/52 = 1/13 = 1/z,$$

$$z = 13$$

Suppose the number of trials is 12 in accordance with (12.21)

$$1 < 12 < 13.$$

If one does not get any queen in 12 drawings, the relative frequency is lower than $P[\mathbf{E_n}^{(Q)}]$

$$0/12 = 0 < 1/13.$$

If one gets one (or more) queen in 12 experiments, the relative frequency is greater than $P[\mathbf{E_n}^{(Q)}]$

$$1/12 > 1/13.$$

In conclusion, the relative frequency $F[\mathbf{E_n}^{(Q)}]$ never matches with $1/13$.

From (12.21) we obtain that $z_{\min} = 2$ and $n_{\min} = 1$, in which case the conclusion (12.23) coincides with (12.7).

12.6.2 In a sense, (12.23) provides the solution to the inverse problem attacked by the theorem of large numbers. While TLN takes n as 'very large', TUBL estimates the threshold of realism, namely, the minimal configuration that is necessary to control probability. Equations (12.21) and (12.22) yield that the number of trials n must exceed z to ensure the testability of P. For example, $P[\mathbf{E}_n^{(e)}] = 10^{-5}$ should refer to no less than 100,000 experiments; if not, it does not have any material meaning.

12.6.3 TUBL is significant on the intellectual plane but has lower importance in the work environment; it could be classified as a technical–mathematical theorem rather than as an aid for professional activities. An individual ordinarily makes a decision on a contingent case and very rarely on a 'small group' of occurrences.

By way of conclusion, TLN, TSN, and TUBL explicate the conceptual origins of the frequentist and subjective conceptualizations. The masters do not pay so much attention to this topic because they are more committed to justifying and defending their theoretical choices (Chapter 3).

Synopsis

a. The elementary event **E** has the property of occurring and does so by means of the specialized elements i, r, and e.

b. Events achieve three main degrees of action and cognition.

c. The compound event $\hat{E}$ occurs due to its sub-events.

d. Probability serves as the measure of the event existence.

e. Probability cannot be systematically tested. Different possibilities of testing generate the frequentist and subjective models.

12.7 Randomness

The concept of randomness must be carefully examined since the definition 8.3 is limited to the necessary characteristics of the indeterminate event. Sufficient conditions cannot be expressed in unified terms in consequence of the incompatible characteristics of physical and mental happenings. It is indispensable to carefully answer the following question:

(**A**) What is randomness?

12.7.1 For the subjectivists and the Bayesians, haphazardness is simply due to prior information that is scarce, inadequate, or improper

$$\mathbf{E}_1^{(pr)} = (\text{Prior information, Belief; Proposition}). \qquad (12.12)$$

'Uncertainty' is the ocean in which epistemic philosophers sail along with psychologists, sociologists, and others (Zhao *et al.*, 2014). Query **A** raises intellectual interest; subjectivists aim to improve the individual's awareness of uncertainty, and Bayesian methods provide effective measures against this real challenge (Section 15.4.2).

12.7.2 Question **A** is very demanding from the physical–ontological stance whose studies fall into two groups.

(**I**) Researchers investigate *specialist and narrow* topics. They look into the *haphazardness of particular systems, machines, games of chance,* etc. For example, Keller (1986) and Diaconis (2007) have studied the unstable equilibrium of a coin thrown into the

Earth's gravitational field; Eichberger (2004) has analyzed the physics of the roulette wheel, and Kapitaniak with colleagues (2012), the kinematics of dice.

(II) The second research thread pursues *broad objectives* and explores the *general characteristics* of randomness. John Venn devoted the entire fifth chapter of *The Logic of Chance* to this argument (Wall, 2005). In the 1920s, von Mises inaugurated a fertile line of inquiries that has advanced up to the present days on the assumptions that *randomness* is:

A. *Objective.* It does not result from human ignorance.

B. *Absolute.* It does not refer to a special context or regards a special case.

C. *Testable.* It is controlled by means of experiments.

Property **A** is typical of long-term experiments in harmony with TLN

$$\mathbf{E}_\infty^{(e)} = ((\mathbf{E}_{11}^{(e)},\ \mathbf{E}_{12}^{(e)},\ \mathbf{E}_{13}^{(e)}, \ldots)). \tag{12.3}$$

Note that the focus on $\mathbf{E}_\infty^{(e)}$ does not constitute a limitation because science requires that experimental observations be replicable and continually reviewed, namely science preferably investigates $\mathbf{E}_\infty^{(e)}$.

Researchers prefer to investigate the outcomes rather than the sub-events of (12.3), and this choice matches with equality (10.8b). For the sake of simplicity, researchers employ two outcomes encoded with ones and zeros

$$e_\infty = ((e_{11},\ e_{12},\ e_{13}, \ldots)) = ((1, 0, 0, 1 \ldots)). \tag{12.24}$$

In principle, $\mathbf{E}_\infty^{(e)}$ can also bring forth a regular sequence of ones and zeros (Section 2.4.7), but point **B** requires dealing with general configurations that do not exhibit a specific pattern. Therefore, researchers assume the complete absence of any periodicity in the output (12.24). They go through patternless distributions of n numbers where every sub-interval $[a, b]$ with $a, b \in [0, n]$ should contain a number of ones and zeros proportional to the length of $[a, b]$ (Kuipers and Niederreiter, 2006)), in other words, a symbol should not be 'denser' in sub-strings than in strings.

12.7.3 Local disorder must occur everywhere, while the distribution of the variable under examination must be uniformly irregular:

"Limiting values [of frequency] must remain the same in all partial sequences which may be selected from the original one in an arbitrary way." (von Mises, 1957)

On the one hand, there is a relative frequency limit, on the other hand, the relative frequency limit does not vary in the operation of selecting the eligible location. Point **C** requires to specify this generic criterion of control and one should select a subsequence of symbols in such a way that the decision to select this subsequence does not depend on the symbols examined. One could ask:

Which are the admissible place selections?

Church (1940) has suggested that the admissible place selections would be the computable ones; that is, every operation is given by a recursive function. This definition has been seen as one of the earliest applications of the Church–Turing thesis.

A string of ones and zeros can be output by a software program executable by a Turing machine. A very complex or disordered string is necessarily created by a 'long' program, whereas a regular string can be computed by a very 'short' program. The *algorithmic complexity theory*, which was developed by Kolmogorov (1998) with Solomonoff and Chaitin, and which remains a cornerstone in the field, deduces three major conclusions from this initial idea.

As first, Kolmogorov fixes the concept of *complexity*. Suppose x is a binary string and p is the computer program, which is written with the given language L and creates x. Various software programs are able to generate x, Kolmogorov states that *the complexity K of x is the shortest length of p*

$$K_L(x) = \min l(p). \qquad (12.25)$$

The second step relates complexity to the idea that the string is *random* if and only if every computer program which produces that string is at least as long as the string itself. Therefore, the *finite string x is random if the complexity $K_L(x)$ is not lower than the length of x*

$$K_L(x) \geq |x|. \qquad (12.26)$$

Finally, Kolmogorov relates the algorithmic randomness to the *compressibility*. He notes that if x has some regularities, it is not random or is partially random, so it can be compressed. As a consequence, the minimal program length significantly reduces in size

$$K_L(x) < |x|. \qquad (12.27)$$

An irregular string cannot be compressed without loss, and the following concept reinforces (12.26). Given the constant $c > 0$ which depends on the computational model, the string x is called *c-incompressible* when

$$K_L(x) \geq |x| - c. \qquad (12.28)$$

12.7.4 Point **C** requires these achievements to undergo a clear criterion for testing, which turns out to be challenging because a sequence can pass one test and fail others. Ideally, an objectively random string should pass all the tests (computationally implemented) and Martin-Löf (1966) suggests an increasingly stringent testing criterion. His idea means to define statistical tests as a countable family of criticality levels. Random strings are those which prove to be random for any algorithm. The collection of c-incompressible strings should coincide with the collection of strings that pass all computationally enumerable statistical tests for randomness.

The scientific community is inclined to regard the Kolmogorov–Martin-Löf criterion as satisfactory (Downey and Hirschfeldt, 2010) because:

• It is applicable to the finite case and the infinite case alike and thus has the needed generality.
• It refers to the string itself and does not require 'external' considerations. In fact, it contains in its definition all that is necessary for its use.

In conclusion, modern studies on objective randomness complement definition 8.3, which merely states the necessary conditions of haphazardness.

12.7.6 *The objective randomness of single events* – The arguments briefly outlined in this section yield another result, which will be used repeatedly in this book.

The specialist approach **I** and the general methods **II** establish the conditions for the randomness of the long-term event that can be concisely expressed this way

$$0 < P[\mathbf{E}_\infty^{(e)}] < 1. \tag{12.29}$$

In consequence of the following definition

$$\mathbf{E}_\infty^{(e)} = ((\mathbf{E}_{11}^{(e)}, \mathbf{E}_{12}^{(e)}, \mathbf{E}_{13}^{(e)}, \dots)). \tag{12.3}$$

And from (12.29) we can formally say that also the single event is random

$$0 < P[\mathbf{E}_{1h}^{(e)}] < 1, \quad h = 1, 2, \dots \tag{12.30}$$

In fact, the long-term event and the sub-events share the same initial conditions while each outcome of (12.24) is unpredictable. The techniques **I** and **II** enable us to recognize the objective randomness of $\mathbf{E}_\infty^{(e)}$ and also of $\mathbf{E}_{1h}^{(e)}$.

The real meaning of (12.30) does not contradict TSN since the theorem assumes $P[\mathbf{E}_1^{(e)}]$ is a precise value and not a generic number placed between zero and one.

The objective and physical value of (12.30) will play a central role in Chapters 16 and 17 of this book that address physical questions.

References

Bernardo J.M. (1996). The concept of exchangeability and its applications, *Far East Journal of Mathematical Sciences*, 4, 111–121.

Beynon M., Curry B., and Morgan P. (2000). The Dempster–Shafer theory of evidence: An alternative approach to multicriteria decision modelling, *Omega*, 28(1), 37–50.

Bulling N. (2014). A survey of multi-agent decision making, *Künstliche Intelligenz*, 28, 147–158.

Carabelli A.M. (1988). *On Keynes's Method* (Palgrave MacMillan, London).

Church A. (1940). On the concept of a random sequence, *Journal of Symbolic Logic*, 5(2), 71–72.

de Finetti B. (1930). Funzione caratteristica di un fenomeno aleatorio, *Atti della R. Accademia Nazionale dei Lincei*, 6(4), 251–299.

de Finetti B. (1933). Classi di numeri aleatori equivalenti, La legge dei grandi numeri nel caso di numeri aleatori equivalenti, Sulla legge di distribuzione dei valori di una successione di numeri equivalenti, (3 papers), *Rendiconti Reale Accademia Nazionale dei Lincei*, 18, 107–110, 203–207, 279–284.

de Finetti B. (1964). Foresight: Its logical laws, its subjective sources, In H.E. Kyburg and H.E. Smokler (eds.), *Studies in Subjective Probability*, 93–158 (Wiley, New York).

de Groot M.H. (1975). *Probability and Statistics* (Addison-Wesley, Boston, MA).

Dempster A.P. (1967). Upper and lower probabilities induced by a multi-valued mapping, *Annals of Mathematical Statistics*, 38, 325–339.

Diaconis P., Holmes S., and Montgomery R. (2007). Dynamical bias in the coin toss, *SIAM Review*, 49(2), 211–235.

Downey R.G. and Hirschfeldt D.R. (2010). *Algorithmic Randomness and Complexity* (Springer, Berlin; New York).

Eichberger J. (2004). *Roulette Physics*, Available at https://www.roulette physics.com/wp-content/uploads/2014/01/Roulette_Physik.pdf.

Galavotti M.C. (ed.) (2009). *Bruno de Finetti: Radical Probabilist* (College Publications, London).

Giere R.N. (1973). Objective single case propensities and the foundations of statistics, In P. Suppes, L. Henkin, A. Joja, and C. Moisil (eds.), *Logic, Methodology and Philosophy of Science*, Vol. 4 (North Holland, Amsterdam).

Jackson E.G. (2020). The relationship between belief and credence, *Philosophy Compass*, 15(6), 1–13.

Kapitaniak M., Strzalko J., Grabski J., and Kapitaniak T. (2012). The three-dimensional dynamics of the die throw, *Chaos*, 22, 047504. Available at https://aip.scitation.org/doi/10.1063/1.4746038.

Keller J.B. (1986). The probability of heads, *The American Mathematical Monthly*, 93, 191–197.

Keynes J.M. (1921). *A Treatise of Probability* (MacMillan & Co., London; New York).

Kolmogorov A. (1998). On tables of random numbers, *Theoretical Computer Science*, 207(2), 387–395.

Kuipers L. and Niederreiter H. (2006). *Uniform Distribution of Sequences* (Dover Publications, Chicago).

Kyburg H.E. (1961). *Probability and the Logic of Rational Belief* (Wesleyan University Press, Middletown, CT).

Martin-Löf P. (1966). The definition of random sequences, *Information and Control*, 9(6), 602–619.

Maximov V.M. (2014). Multidimensional and abstract probability, *Proceedings of the Steklov Institute of Mathematics*, 287, 174–201.

Morris T.V. (1986). Pascalian wagering, *Canadian Journal of Philosophy*, 16, 437–54.

Popper K. (1963). *Conjectures and Refutations* (Routledge and Keagan Paul, London).

Ramsey F.P. (1931). Truth and probability, In R.B. Braithwaite (ed.), *Foundations of Mathematics and other Logical Essays* (Kegan, Paul, Trench, Trubner & Co., New York).

Révész P. (1967). *The Laws of Large Numbers* (Academic Press, Cambridge, MA).

Rocchi P. (2014). *Janus-faced Probability* (Springer, Berlin; New York).

Rocchi P. (2017). Four foundational questions in probability theory and statistics, *Physics Essays*, 30(3), 314–321.

von Mises R. (1957). *Probability, Statistics and Truth* (MacMillan Co., London; New York).

Wall B.E. (2005). Causation, randomness, and pseudo-randomness in John Venn's Logic of Chance, *History and Philosophy of Logic*, 26(4), 299–319.

Walley P. (1987). Belief-function representations of statistical evidence, *Annals of Statistics*, 10, 741–761.

Zhao J., Hahn U., and Osherson D. (2014). Perception and identification of random events, *Journal of Experimental Psychology: Human Perception and Performance*, 40(4), 1358–1371.

Chapter 13

Consequence of Probability Testing

The theorems of the previous chapter look to be of little significance or even trivial from the mathematical stance. The reader might be induced to underestimate their weight which instead provides precise answers to relevant and long-standing issues. Let us go through four conclusions derived from the theorems just mentioned.

13.1 Precise Hypotheses

This book introduces $P(\mathbf{E})$ as a *theoretical parameter* that is consistent with the perspective of mathematicians who treat probability in the abstract (Segal, 1954). Instead, in working environments, $P(\mathbf{E_n})$ is subject to the following restrictions:

$$n \to \infty \qquad (13.1a)$$

$$1 \leq n < z. \qquad (13.1b)$$

The constraints (13.1a) and (13.1b) govern the control of P in the world; they determine the realism and unrealism of the probability values and make explicit the areas accessed by the frequentist and subjectivist approaches.

This book often refers to $n = 1$ instead of (13.1b) and the reader arguably may ask:

What about the intermediate cases placed between $(n \to \infty)$ and $(n = 1)$?

Prevailing experience shows that scientists who search for general explanations – i.e., the laws of nature – operate with large populations and classical methods. The Bayesians are mostly concerned with decision-making and individual predictions. In summary, professional practice shows that experts spontaneously tend to operate in harmony with $(n \to \infty)$ and $(n = 1)$, while the intermediate values are somewhat distant from the goals of statistical projects or anyway draw less interest.

13.2 Complementary Models

The central importance of testing in science entails that frequentist and subjectivist models are the most significant in real-life applications. The following theorem, descending from the previous section, proves that they are not incompatible as usually credited (Bergmann, 1945; Bunge, 1981).

Theorem 13.1. *Theorem of Logical Compatibility (TLC)*
The frequentist and subjective probabilities are disjoint

$$P[\mathbf{E}^{(e)}] = P[\mathbf{E}_{\infty}^{(e)}] \text{ OR } P[\mathbf{E}_{1}^{(e)}]. \tag{13.2}$$

Proof. The number n belongs to the separate intervals (13.1a) and (13.1b), so a scientist calculates either $P[\mathbf{E}_{\infty}^{(e)}]$ or $P[\mathbf{E}_{1}^{(e)}]$. The two distant contexts do not imply any logical incongruity.

The frequentist and subjective models provide alternative mathematical supports even though they refer to the same case.

Corollary 13.1. *Corollary of Separation*
When

$$\mathbf{E}_{\infty}^{(e)} = ((\mathbf{E}_{11}^{(e)}, \mathbf{E}_{12}^{(e)}, \mathbf{E}_{13}^{(e)}, \ldots)), \tag{12.3}$$

the probabilities $P[\mathbf{E}_{\infty}^{(e)}]$ and $P[\mathbf{E}_{1h}^{(e)}]$ ($h = $ any of $1, 2, 3 \ldots$) have distinct testing properties

Proof. TLN and TSN regulate the control of the left-hand side and right-hand side of (12.3), respectively. The events $\mathbf{E}_{\infty}^{(e)}$ and $\mathbf{E}_{1h}^{(e)}$ (h = any of $1, 2, 3 \ldots$) belong to disjoined domains of application in agreement with TLC.

TLN ensures the testability of $P[\mathbf{E}_{\infty}^{(e)}]$, whereas TSN denies $P[\mathbf{E}_{1h}^{(e)}]$ can be tested; therefore, there is no continuity between the management of the two probabilities even though they belong to a common overall fact. The same can be said for $P(e_{\infty})$ and $P(e_{1h})$ due to (10.8b).

13.3 Guidelines for Statisticians

The frequentist and the subjective interpretations provide the most significant models from the application viewpoint and suggest the guidelines for statisticians who begin a work plan (Section 5.2). The criteria that identify frequentist and subjective probabilities also teach how to solve the problem of selecting the most suitable statistical inference (Johannesson, 2020; Hackenberger, 2019). A statistician should simply use (13.1a) and (13.1b) in a project.

Criteria for Discernment (CD)

☐ *If one deals with a large number of cases* ($n \to \infty$), *then one resorts to frequentism and employs classical statistics.*

☐ *If one deals with a single case* ($n = 1$), *then one resorts to subjectivism and employs Bayesian statistics.* (13.3)

A statistician should no longer choose and develop a statistical method at will, exploit personal opinions, or invoke ethereal principles. The two rules revolve around n that is a solid and well-controllable parameter, so they cannot raise doubts or uncertainties.

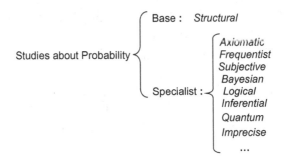

Figure 13.1. Classification of the probability theories.

13.4 Many Directions

The current structural theory spells out the fundamentals of probability and statistics through the phenomenological approach, which deduces the properties of $P(\mathbf{E})$ from $\mathbf{E}$.

In particular, the ability to occur is the essential quality of events; and $\mathbf{E}$ is the causal determination of $P(\mathbf{E})$ which measures this quality. For this reason, different meanings of probability come from different structures; $P(\mathbf{E})$ is multifold in consequence of the varieties of $\mathbf{E}$, and the present work definitively denies the criterion of uniqueness which claims that only one model should suffice to interpret probability (Chapter 4).

The mathematical community investigates the immense variety of random occurrences and divides this domain of knowledge into specialized fields: frequentists pay attention to collectives, subjectivists and Bayesians to individual cases, quantum scientists to the microscopic world, Keynes to rational belief, and so on. Axiomatic theory provides agile support at the abstract level, and other purely mathematical theories may be developed in the future. Each school explores a particular area or pursues a specific scope; each one develops its own perspective (Figure 13.1).

The numerous models of P are no longer in conflict because they derive from precise hypotheses and constraints, which this framework makes explicit. The various achievements could be compared to advanced courses, while the tenets, deduced here, give material to the base course that illustrates the shared conceptual roots.

References

Bergmann G. (1945). Frequencies, probabilities, and positivism, *Philosophy and Phenomenological Research*, 6(1), 26–44.

Bunge M. (1981). Four concepts of probability, *Applied Mathematical Modelling*, 5, 306–312.

Hackenberger B.K. (2019). Bayes or not Bayes, is this the question? *Croatian Medical Journal*, 60(1), 50–52.

Johannesson E. (2020). Classical versus Bayesian statistics, *Philosophy of Science*, 87(2), 302–318.

Segal I.E. (1954). Abstract probability spaces and a theorem of Kolmogorov, *American Journal of Mathematics*, 76(3), 721–732.

Chapter 14

How Discrepant Features Cohabit

Thinkers argue about opposed and sometime irreconcilable aspects of haphazardness that can emerge even in a single problem. The following pages are devoted to three distinct arguments and demonstrate how heterogeneous features cohabit without conflicts.

14.1 Fading Critiques

The censurers of von Mises claim:

(1) The relative frequency should come from infinite repetitions which are merely hypothetical, and in fact, only a finite number of experiments can be performed.

(2) The relative frequency slightly varies every time it is measured. Different frequencies will appear in different series of trials that instead should be the same every time.

(3) Suppose we acknowledge that probability can be evaluated with some error of measurement. In that case, we meet a defect of logical circularity since the error of measurement should be expressed in terms of probability.

The probability $P[\mathbf{E}_\infty]$ cannot be recognized as the 'universal' model presumed by von Mises and his followers because it is subjected to the constraint $(n \to \infty)$. Various writers agree that the frequentist framework does not have the expected generality (Hajek, 2009; Lewis, 1986; Jaynes, 1984), hence the adverse notes **(1)** and **(2)** lose meaning. The fluctuations of the frequency can be handled by the criteria typical of scientific measurement processes, and like-lihood methods do not create the fallacy of circularity **(3)**.

Severe objections have been raised against the subjective–Bayesian framework (Smith, 1984; Gelman, 2008), such as the following:

(a) Subjective probabilities differ from person to person and contain a high degree of personal bias. The variety of an individual's convictions affects his or her assessment of probability. For instance, a certain amount of money involved in an aleatory occurrence can be significant for one person but marginal for another.

(b) The pattern of consistent bets appears to be questionable. For instance, gamblers seek the risk in gambling, whereas the customer of an insurance company tends to minimize the risk.

(c) The betting scheme turns out to be strange in the scientific and engineering sectors. For instance, a quantum scientist who estimates the position of an electron should play the role akin to a gambler.

(d) Testing is a key criterion for the scientific method, but the subjective probability is alien to any experimental validation. This turns out to be repellent to science and technology, which handle objective situations and strive for numerical results that are independent of human influence.

Objection **(a)** can be addressed to any epistemic probability and is essentially bound to the nature of mental events. Subjectivists and Bayesians conducted an intelligent plan of recovery, and annotations **(b)** and **(c)** are nothing more than the price paid to reuse $P[\mathbf{E_1}]$ that otherwise should be discarded. TSN proves that the probability of the single experiment is untestable and **(d)** recognizes this unavoidable limit.

In summary, the theorems of large numbers and the single number prove that neither frequentism nor subjectivism provides the 'authentic' concept of probability. For this reason, the listed criticisms remain true but have completely different imports and tones inside the present framework. The tough judgements lose some of their force in the light of the theorems presented here that explain the origins and justify the limits of the principal probabilistic models.

14.2 Diverging Routes, a Unique Starting Point

Classical and Bayesian statistics show discrepant conceptual sides that call for insights.

14.2.1 The former makes propositions about a population using data drawn from that population with some form of sampling in harmony with the assumption $(n \rightarrow \infty)$ (Barnett, 2009). It is feasible to repeat an experiment several times with some stopping criterion, and data that are observed under similar circumstances, differ. Randomness is associated with the variations in replicated observations, while the studied parameters are not aleatory variables. Experts can employ an assortment of techniques to tune the parameters of the distribution which in principle is posited as fixed. Traditional inference procedures resort to significance tests, while some concepts – e.g., p-values, confidence intervals, statistical power, etc. (Greenland *et al.*, 2016) – come from the idea that probability is proportional to the number of results in the long run (theorem of large numbers).

The input data could be classified as the 'premises' of the logical process, while the statistical achievements could be labeled as the 'conclusion'. By definition, deductive reasoning comes from one or more premises to logically reach a conclusion, so classical statistics complies with the *deductive* or *inferential logic*.

14.2.2 For Bayesians, probabilities are numerical values for making decisions based on incomplete information. Specific approaches differ in emphasis – subjective or objective, personalist or logical, etc. – yet they share several mathematical equations.

A Bayesian analysis begins with the quantification of subjective priors derived from the statistician's existing state of knowledge, beliefs, and assumptions, which are taken just as they are. The probability distribution over a parameter (or set of parameters) expresses personal knowledge about the true value of that parameter (or set of parameters). The analysis proceeds based on Bayes's theorem which

Table 14.1. Essential traits of classical and Bayesian statistics.

	Classical	Bayesian
Observed data	Random	Fixed
Model parameters	Fixed	Random
Inference	Deductive	Inductive

computes the posterior using the prior and likelihood known for all hypotheses. The Bayesian school provides a standard set of procedures to perform calculations. One might conclude that the Bayesians believe the parameter is an uncertain quantity and subject to probabilistic description, while observations are fixed since the Bayesians do not have any doubt about the reliability of collected data.

Inductive reasoning attains a conclusion by extrapolating from specific cases to general statements, although residual uncertainty cannot be discarded. The Bayesian course starting with specific knowledge and moving to the posterior distribution is associated with *inductive logic*.

Classical statisticians posit the parameter under scrutiny as fixed but unknown and the data are treated as variables. Parameters are estimated using samples from a population; conversely, the Bayesian approach treats the parameters of interest as random variables (Table 14.1).

14.2.3 Paper (Rocchi *et al.*, 2010) reviews a sample of books printed between the years 1941 and 2008 and has the purpose of analyzing the diffusion of the 'dualist' literature. Bibliographic data show how the inclination to accept both the statistical methods is gaining space by time passing. Especially, educational texts seek to offer a complete account of both classical and Bayesian statistics. At the same time, the sampled writers tend to place the conflicting principles typical of the two contexts in the background. They neglect the conceptual differences and present the statistical methods as a sort of 'cooking recipes collection'. As one prepares his favorite dish in the kitchen,

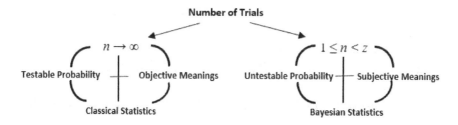

Figure 14.1. Conceptual map of frequentism and subjectivism.

so the writers give the impression that a statistical 'recipe' may be chosen at personal convenience.

The present study places all the topics inside a comprehensive frame. Every notion pertains to a precise conceptual web and has links with all the remaining notions but has nothing to do with the elements of the other web. We face two intellectual areas, which turn out to be logically congruent and look like distinct 'worlds', that the number of trials bridges (Figure 14.1) on the conceptual and practical plane too, in fact, the criteria for selection (13.3) center on n, which is a well-controllable parameter.

This research is close to the studies which endeavor to foster the cultural progress of statistics. They contribute to the maturation of this discipline (Sowey and Petocz, 2017; Bandyopadhyay and Forster, 2011) and are intended to formulate a unified basis of thought for statisticians (Stigler, 2016).

14.3 Global Order and Local Disorder

The *causality principle*, a cornerstone in logic and science, refers to the way one entity – material or abstract – produces another entity. In broad strokes, this principle recognizes the *causes* that under certain conditions give rise to something else, called the *effect*.

Two rival philosophical schools are entirely interlinked with the principle of causality. *Determinism* and *indeterminism* are doctrines that set opposed philosophical visions (Kraal, 2013). The first maintains that all events are governed completely by previously existing causes; the behaviors of animate and inanimate entities have precise origins and can be predicted at least in principle. Free will is an illusion, and humans are guided by internal or external forces over which they have no control. The search for the 'first cause' becomes a dominant argument for deterministic thinkers.

The opposing school of thought is grounded on the idea that the phenomena in the world have no precise origin, nothing is certain, and the entire universe is intrinsically accidental. The followers argue about the notions of haphazardness, indeterminacy, and fuzziness, taking inspiration from science and technology.

14.3.1 In the present structural theory, the components i, r, and e make a chain where the third element depends on the second, and in turn this relies on the first. The cause/effect mechanisms appear to be evident also for the random event

$$\mathbf{E}^{(e)} = (i, r; e). \tag{8.6}$$

Causality governs $\mathbf{E}^{(e)}$, even if there are no tight rules. The semicolon indicates a relation that can fail but anyway is working. Chance is not opposed to *general causation*, it is opposed to *stringent causation*. The triad (8.6) provides an analytical answer to the long-vexed dilemma about chance and causality (Section 2.4). Conversely, the schools of thought mentioned above guide listeners toward 'metaphysical' conclusions that are fundamentally contrary to each other. They lean toward a binary view encompassing 'determinism' as the opposite to 'indeterminism' just as 'black' is the opposite of 'white,' or 'bad' is the opposite of 'good.' They overlook the 'mid tones', and in a way reject the concept of probability that is the meter measuring 'intermediate' situations (Section 10.1).

14.3.2 Radical beliefs spawn profound doubts such as the following question about the LLN, already mentioned in Section 2.3, which seems to be an unresolvable conundrum:

$\mathcal{A}$) Why and how can outcomes that are considered independent, irregular, and in isolation be subjected to a common rule?

An idea can be found in the *general system theory* (GST), inaugurated by the biologist Ludwig von Bertalanffy (1968) in the 1950s in order to explore complex entities. Systems are defined as functionally related sets of interacting and interdependent elements which work together for the whole to function successfully (Rousseau *et al.*, 2018). *System-thinking* takes two directions: it rises above the separate components and looks at the whole; on the other hand, it breaks down the system into each part and checks every element (Esteves, 2020). The first intellectual approach, called *holistic thinking*, is developed alongside the second approach, called *reductionist thinking* (de Rosnay, 1979).

14.3.3 Query $\mathcal{A}$ calls to mind some questions typically raised by system-thinking:

- Why can particles with *many degrees of freedom* have together a *few degrees of freedom*?
- How can *inanimate* components make a *living* being?
- How can *unaware* agents (e.g., termites) make an *efficient* organization?
- What separates *living* matter from *non-living* matter?

GST experts argue about these kinds of questions; they discuss whether the properties of the system are explanatorily reducible to the properties of its parts, and at present, they agree in saying that the behavior of the system is inherently unpredictable from the components. It is impossible to trace the constitutive, logical, and causal chains that go from the least elements to the overall system (Minati and Collen, 1997). The macroscopic properties that exhibit blatant differences from the individual constituents raise non-trivial difficulties in system management, design, control, etc. and are often labeled as the *problem of emergence*.

14.3.4 Let us look into the problem of emergence using the *structure of levels* (SL) that describes the system at different grades of granularity (Debit *et al.*, 2001; Poli, 1998).

The standard symbol of equality (=) shows how the first term equals the second term, conversely, SL says that one level differs from another one. For instance, (8.6) holds that the whole is made of parts, instead SL stresses how the whole and the parts are distinct entities

$$\text{Level 1} \quad \mathbf{E}^{(e)}$$
$$\text{Level 2} \quad (i, r; e). \tag{14.1}$$

It may be said that (14.1) illustrates a perspective that, while not contradicting (8.6), emphasizes the differences between the levels.

Let us examine the long-term event called in question by $\mathcal{A}$. The corollary of separation (Section 13.2) demonstrates the divide between the left-hand and right-hand sides of (12.3) and is consistent with the following SL:

$$\text{Level 1} \quad \mathbf{E}_{\infty}^{(e)}$$
$$\text{Level 2} \quad ((\mathbf{E}_{11}^{(e)}, \ \mathbf{E}_{12}^{(e)}, \ \mathbf{E}_{13}^{(e)}, \dots)). \tag{14.2}$$

The layers of (14.2) show different behaviors, in fact, the probability of each individual occurrence is out of control, whereas the empirical frequency corroborates $P[\mathbf{E}_{\infty}^{(e)}]$. The layers underscore these far different characteristics in conformity with the more general problem of emergence.

14.3.5 The theorem of the complete structure shows how $\hat{\mathbf{E}}^{[N)}$ surely supplies the outcome e, whereas each sub-event has an unpredictable conduct (Section 9.2). We obtain the following structure whose levels 2 and 3 present the determinate triads, and level 4 exhibits the random triads

$$\text{Level 1} \quad \hat{\mathbf{E}}^{[N)}$$
$$\text{Level 2} \quad (i, \, r - e)$$
$$\text{Level 3} \quad [(i, \, r) - (e_1 \, \text{OR} \, e_2 \dots \text{OR} \, e_N)] \tag{14.3}$$
$$\text{Level 4} \quad [(i, \, r; e_1) \, \text{OR} \, (i, \, r; \, e_2) \, \text{OR} \dots \text{OR} \, (i, \, r; \, e_N)].$$

Different elements cohabit but do not raise logical conflicts since they belong to different layers. Aleatory events share the behavior of the complex entities investigated by GST; they exhibit features that the parts do not have on their own.

14.3.6 The whole and the parts of (14.3) are variant, namely, they have identical initial conditions and only the outcomes of the sub-events vary. The structure proves that the latter have little weight since they belong to the lower layer (Rocchi, 2003), and this remark is consistent with the fact that they make a definite set (the sample space). Hence, they cause fluctuations and not radical overthrows.

The present conclusions match with the writers claiming that *partial precision* characterizes random physical phenomena. Poincaré, Cournot, Poisson, and others relate the haphazardness to *microscopic elements, inexact modes, faulty* or *incomplete functions, generic conditions, partial responses*, etc.

> "These events are governed partly by constant factors and partly by variable factors whose variations are irregular and do not cause any systematic change in a definitive direction." (Poisson, 1837)

Minute factors, which belong to the lower layers of (14.2), bring forth *modest effects that consist of oscillations, or even deviations, but do not cause total chaos*. Local irregularities and global regularities do not conflict but cooperate to reach common objectives and provide the complete answer to question $\mathcal{A}$.

14.3.7 The subjectivists, the Bayesians, and others relate an uncertain occurrence to *defective initial cognition* and not *absolute ignorance*. A wealth of prior information – never mind whether it is scarce, vague, or imprecise – supports the assessment of $P[\mathbf{E}_1^{(pr)}]$. Burdzy (2009) sets off how paradoxically a subjectivist cannot study an event that has nothing in common with anything else.

In accordance with the cognitive perspective, human unawareness is a rather ordinary condition; it may even be regulated by an individual's will. Sometimes scientists explicitly delimit their own knowledge. For example, *descriptive statistics* seeks the shared features of phenomena and provides summaries that neglect the particulars. Surveys, polls, inquiries, etc. watch the intended happening 'from above' and systematically disregard minute elements.

14.3.8 Concluding, the present structural theory demonstrates that the limits of frequentism and subjectivism fade away since they are no longer the presumed 'authentic' probabilities.

Inductive reasoning is conceived as the way going from the specific to the general, while deductive reasoning goes from general premises to specific conclusions, hence, classical and Bayesian statistics apply the two fundamental research modes. They do not contrast, conversely it may be said that the former and the latter provide the complete set of means and methods for empirical research.

Finally, aleatory events are neither completely disordered nor totally obscure. They exhibit precise traits together with uncontrollable details. These discordant features coexist at different grades of magnification.

14.3.9 Conservative thinkers – Rivers of ink will flow until probabilists continue to adhere to the minimalist philosophy and live in their own world. They will waste time indefinitely as long as they do not take innovative intellectual inputs into consideration. They should heed to the progress registered in collateral research areas such as the theory of events, the general system theory, semiotics, computer science, and so forth.

Researchers prefer to look into the metaphysical diversities between determinism and indeterminism; chance is conceived as opposed to causation; each school of thought finds dramatic defects in its opponents; aleatory events are considered wholly unpredictable; statistical methods cannot be compared, and solved problems become paradoxical. It is not an exaggeration to conclude that the questions are dramatized for no reason.

On the other hand, we find those who accept that the event and the result are the same thing, judge the logical incongruities between the schools of thought as useless philosophical controversies, choose the statistical method at personal will as if it were like choosing which hat to wear, and sweep other problems under the rug.

References

Bandyopadhyay P. and Forster M. (eds.) (2011). *Handbook for the Philosophy of Science: Philosophy of Statistics* (Elsevier, London).
Barnett V. (2009). *Comparative Statistical Inference*, 3rd edition (Wiley, New York).

Burdzy K. (2009). *Search for Certainty: On the Clash of Science and Philosophy of Probability* (World Scientific Pub. Co., Singapore).

Debit N., Garbey M., Houston, Hoppe R., Périaux J., Keyes D., and Kuznetsov Y. (2001). Domain decomposition methods in science and engineering, *Proceedings of the 13th International Conference on Domain Decomposition Methods*, Available at http://www.ddm.org/DD13/Proc-13.php.

de Rosnay J. (1979). *The Macroscope. A New World Scientific System* (Harper & Row, New York).

Esteves M.J. (2020). *Systems Thinking: The New Paradigm of Science* (Simplíssimo).

Gelman A. (2008). Objections to Bayesian statistics, *Bayesian Analysis*, 3(3), 445–450.

Greenland S., Senn S.J., Rothman K.J., Carlin J.B., Poole C., Goodman S.N., and Altman D.G. (2016). Statistical tests, p-values, confidence intervals, and power: A guide to misinterpretations, *European Journal of Epidemiology*, 31, 337–350.

Hajek A. (2009). Fifteen arguments against hypothetical frequentism, *Erkenntnis*, 70(2), 211–235.

Jaynes E.T. (1984). The intuitive inadequacy of classical statistics, *Epistemologia*, VII, 43–74.

Kraal A. (2013). Determinism and indeterminism, In A.L.C. Runehov and L. Oviedo (eds.), *Encyclopedia of Sciences and Religions* (Springer, Berlin; New York).

Lewis D. (1986). *Philosophical Papers*, Vol. II (Oxford University Press, Oxford).

Minati G. and Collen A. (1997). *Introduction to Systemics* (Eagleye Books Intleek, San Francisco, CA).

Poisson S.D. (1837). *Recherche sur la Probabilité de Jugements en Matière Criminelle et en Matière Civile* (Bachelier, Paris).

Poli R. (1998). Levels, *Axiomathes*, 1–2, 197–211.

Rocchi P. (2003). *The Structural Theory of Probability; New Ideas from Computer Science on the Ancient Problem of Probability Interpretation* (Kluwer/Plenum, New York).

Rocchi P., Pandolfi S., and Rocchi L. (2010). Classical and Bayesian statistics: A survey upon the dualist production, *International Journal of Pure and Applied Mathematics*, 58(3), 255–280.

Rousseau D., Wilby J., Billingham J., and Blachfellner S. (2018). *General Systemology: Transdisciplinarity for Discovery, Insight and Innovation* (Springer, Berlin; New York).

Smith A.F.M. (1984). Present position and potential developments: Some personal views Bayesian statistics, *Journal of the Royal Statistical Society: Series A*, 147(2), 278–292.

Sowey E. and Petocz P. (2017). *A Panorama of Statistics: Perspectives, Puzzles and Paradoxes in Statistics* (Wiley, New York).

Stigler S.M. (2016). *The Seven Pillars of Statistical Wisdom* (Harvard University Press, Cambridge, MA).

von Bertalanffy L. (1968). *General System Theory* (George Braziller, New York).

Chapter 15

Conditional Probability

Conditioning is a noteworthy topic and this chapter reviews some significant questions in conformity with the phenomenological logic typical of this framework.

15.1 Ideal and Real Situations

Speaking in general, theorists pay attention to perfect phenomena that, by definition, are not polluted by extraneous elements or influenced by strange factors. For example, chemistry begins with pure and unmixed substances; electronics analyzes circuits that do not dissipate energy; and the first law of mechanics describes the motion of a body unaffected by attrition. Even the lessons of probability begin with perfect mechanisms consisting of fair coins, perfectly balanced roulette wheels, and so on.

However, practitioners face a different landscape than theorists. They often meet spurious factors and intrusions, while ideal cases prove to be rarer. Gambling is an ideal process in textbooks, whereas cheating and fraud are constant threats in the real environment. For example, systematic countermeasures are needed in a casino to guarantee that dishonest patrons do not influence the games in their favor. Security officers patrol and inspect properties to protect against illegal activities. Audio and visual surveillance equipment assist the security team of the casinos and monitor all rooms and players. This is just to conclude that a large number of factors act on random events and conditional probabilities qualify for very common phenomena.

No doubt that unconditional probability is an ideal measure and regards special situations. It can be said that probability theory would neither support science and professional activities nor be a meaningful field of study without the concept of 'conditioning'. This agrees perfectly with Hájek (2003) who claims that conditional probability cannot be taken to be a purely technical–mathematical element. It is an essential concept whereas unconditional actions would be seen as idealized and didactically oriented.

Early in the 1940s, Koopman (1941) and Copeland (1941) investigated the concept of conditional event with the intent of providing a more solid foundation for this topic. The Bayesian school has enhanced and refined our knowledge and has made us more aware of the complexity of the current topic.

15.2 Introductory Descriptions

The *conditional probability* $P(B|A)$ is usually presented in terms of the following kind:

1] *"The probability of the event B if A occurred"*.
2] *"The conditional probability of the event B under the condition A"*.
3] *"The conditional probability of B knowing that A happened"*.
4] *"The conditional probability of B given that A is true"*.

These statements often introduce the following formula where A and B are subsets of the sample space Ω:

$$P(B \mid A) = \frac{P(A \cap B)}{P(A)}, \quad P(A) > 0. \tag{15.1}$$

The verbal definitions considerably differ from each other because 1] and 2] present physical facts; sentences 3] and 4] refer to human cognition. Equation (15.1) gives the impression of unifying the present area, yet this is not exactly true.

15.3 Conditioning and Independence

Dictionaries say that '*conditioning*' stands for "a significant influence or determination of something", and this perfectly matches with the dynamic nature of events that are able to modify one another.

This book centers on the entities that have the ability to exist (property 7.1) and, therefore, can be influenced by various factors. In particular, we analyze how the general property that $P[\mathbf{E}^{(b)}]$ quantifies, can be affected by $\mathbf{E}^{(a)}$, which either reduces or enhances the possibilities of $\mathbf{E}^{(b)}$ to happen. The *conditional relation* (Section 7.1.2) formalizes the influence of the event $\mathbf{E}^{(a)}$ over the occurrence of $\mathbf{E}^{(b)}$

$$[\mathbf{E}^{(b)} \mid \mathbf{E}^{(a)}] \tag{15.2}$$

while $P[\mathbf{E}^{(b)} \mid \mathbf{E}^{(a)}]$ qualifies it. Property 10.2 ensures the use of results to calculate conditional probability: $P(b \mid a) = P[\mathbf{E}^{(b)} \mid \mathbf{E}^{(a)}]$, that is the standard method in the current literature.

15.3.1 The terms '*conditional*' and '*independent*' are antithetical and so close that they look like two sides of the same coin. Authors explain the first using the second or vice versa. The two concepts are usually introduced through a 'black or white' criterion. The event $\mathbf{E}^{(b)}$ is *independent* of $\mathbf{E}^{(a)}$ if the *conditional probability* of $\mathbf{E}^{(b)}$ given $\mathbf{E}^{(a)}$ is the same as the *unconditional probability* of $\mathbf{E}^{(b)}$

$$P[\mathbf{E}^{(b)} | \mathbf{E}^{(a)}] = P[\mathbf{E}^{(b)}]. \tag{15.3}$$

This criterion applies, for instance, to *independent and identically distributed* (i.i.d.) variables where the occurrence of the generic value e_j is not influenced by any previous element e_k

$$P(e_j \mid e_k) = P(e_j), k < j.$$

15.3.2 Authors also put forward the definition of independence which comes from the multiplication rule, namely, the probabilities $P(A)$ and $P(B)$ are independent when

$$P(A) \bullet P(B) = P(A \cap B). \tag{15.4}$$

von Mises (1963) criticizes this statement using six random values as an example, and poses the question

> "What is the meaning of the statement that, for a given distribution, the event $A = (2, 3, 4)$ and $C = (2, 5)$ are independent, while the events $(2,3,4)$ and $(1,2,5)$ are dependent or events $(1,6)$ and $(2,3,4)$ are dependent?"

He concludes that the multiplication law can be satisfied due to "purely numerical accidents."

15.3.3 The presumed 'general' definition of conditional probability calculates subsets

$$P(B\,|\,A) = \frac{P(A \cap B)}{P(A)}, \quad P(A) > 0. \tag{15.1}$$

Subsets represent special types of outcomes, furthermore $P(B\,|\,A)$ depends on the size of $(A \cap B)$ (Section 15.6); in conclusion (15.1) assesses a particular case.

In conclusion, equations (15.1), (15.3), and (15.4) qualify particular conditioned events, including accidental combinations, and they are stated in the abstract, apart from the frequentist and subjectivist contexts. In order to overcome these partial views, it is necessary to take a broad perspective, consistent with the current phenomenological approach.

15.4 Ubiquitous Factors

Events are material and mental, expression (15.2) subsumes the physical descriptions **1]** and **2]**, and even the sentences **3]** and **4]** (Section 15.2). Let us look into the ontological and epistemic domains of application.

15.4.1 *Physical factors* **(1)** – When $\mathbf{E}^{(a)}$ takes place in the world and comes before $\mathbf{E}^{(b)}$ in the time scale, it can act upon the tangible fact $\mathbf{E}^{(b)}$ and modify its ability to happen.

Given $\mathbf{E}^{(b)}$, an infinite variety of regular or accidental actions can change its existence. Medicine, engineering, chemistry and other sciences look into determinants which sometimes act in a manner that one cannot imagine. Those fields of study teach that harmful aleatory elements sometimes are lurking in the background, and sooner or later change the system under scrutiny. Random perturbations can even result in a chain of effects that can neither be accurately forecasted nor be analyzed from its earliest origin.

15.4.2 *Cognitive factors*(2) – The epistemic realm presents a far different landscape. 'Prior information' is the systematic factor making any subjective probability whatever to be conditional (Chapter 12). De Finetti (1974) claims

> "Every prevision, and, in particular, every evaluation of probability, is conditional; not only on the mentality or psychology of the individual involved, at the time in question, but also, and especially, on the state of information in which he finds himself at that moment."

The triad clearly exhibits the first component that affects the entire intellectual event and makes the subjective probability inherently conditional

$$\mathbf{E}_1^{(pr)} = (\text{Prior information, Belief; Proposition}). \qquad (12.12)$$

Because of the central importance of 'Prior information', Bayesian statistics teaches how to deepen this theme.

Initial, raw information changes in (12.12) when updated data are available. Improved cognition can be combined with pre-existing information leading to updating of $P[\mathbf{E}_1^{(pr)}]$. The interplay of initial information, data, awareness, perception, insight, and other informational factors form the backbone of Bayesian inference where the *Bayes theorem* dominates as the fundamental tool to better probability independently from the application context.

It begins with the idea that conditional relationships between two mental events can be symmetric. More precisely, if information $\mathbf{E}^{(prA)})$ can affect $P[\mathbf{E}^{(prB)}]$, then information $\mathbf{E}^{(prB)}$ can also influence $P[\mathbf{E}^{(prA)}]$

$$P\left[\mathbf{E}^{(prB)} \mid \mathbf{E}^{(prA)}\right] \bullet P\left[\mathbf{E}^{(prA)}\right] = P[\mathbf{E}^{(prA)} \mid \mathbf{E}^{(prB)}] \bullet P[\mathbf{E}^{(prB)}]. \qquad (15.5)$$

The Bayesians reinterpret (15.5) and obtain the *posterior probability* $P[\mathbf{E}^{(prA)} \mid E^{(prB)}]$ from the *prior probability* $P[\mathbf{E}^{(prA)}]$

$$P\left[\mathbf{E}^{(prA)} \mid \mathbf{E}^{(prB)}\right] = P[\mathbf{E}^{(prA)}] \frac{P\left[\mathbf{E}^{(prB)} \mid \mathbf{E}^{(prA)}\right]}{P\left[\mathbf{E}^{(prB)}\right]}. \tag{15.6}$$

where $P[\mathbf{E}^{(prB)} \mid E^{(prA)}]$ is called the *likelihood* of $\mathbf{E}^{(prB)}$ given $\mathbf{E}^{(prA)}$, and $P[\mathbf{E}^{(prB)}]$ is the evidence. In other words, we have

$$Posterior = Prior\left(\frac{Likelihood}{Evidence}\right). \tag{15.7}$$

Bayes's theorem gives the probability of an event based on new information that is, or may be, related to that event. The formula can also be used to determine how the probability of an occurring event may be affected by hypothetical new information, supposing the new information will turn out to be true.

The Bayes theorem demands that we think of conditional probabilities as depending on their *conditional inverse*. It works as a process for obtaining posterior distributions or predictions based on a range of assumptions about both prior distributions and likelihoods. It may be said that Bayes's theorem is a way of converting prior beliefs into posterior probabilities using new data. The new data show up in the likelihood and evidence. In fact, the ratio of likelihood/evidence is a way of normalizing expectations about data based on their marginal probability.

15.5 Frequentist and Subjectivist Contexts

Let us examine conditional probability in relation to the frequentist and subjective environments.

15.5.1 The theorems of large numbers and a single number lead to the following points:

(a) If $(n \to \infty)$, then the conditional probability is a testable and objective quantity.
(b) If $(n = 1)$, then the conditional probability is out of control.

Factors **(1)** and **(2)** regard the probabilities **(a)** and **(b)**, however, the relations between the two pairs are not linear and there is a certain interplay. Factors **(1)** and **(2)** do not correspond to the ontic and epistemic domains strictly speaking. Let us analyze this misalignment.

15.5.2 *Long-term event* (a) – Repetitions imply the interference between $\mathbf{E}^{(b)}$ and $\mathbf{E}^{(a)}$ reiterates again and again, in such a way that $\mathbf{E}_\infty^{(a)}$ affects the overall phenomenon and the meaning of $P[\mathbf{E}_\infty^{(b)} \mid E_\infty^{(a)}]$ is real.

Example. Men and women who reside in a luxurious apartment complex enjoy the gym facilities. The contingency table presents the following survey:

	Men	Women	**Total**
Used gym	65	145	210
Did not used gym	105	35	140
Total	170	180	**350**

One wonders:

a) What is the probability of picking a woman and that the woman uses the gym facilities?
b) What is the probability that a person chosen at random used the gym given that the person was a woman?

The queries regard all the women who live in the apartment complex and used the facilities without limits of time, thus the problem regards two long-term events.

Question a) considers the female gender and the use of the gym facilities that are concurrent. Men and women make two groups; in parallel, there are two groups which use and do not use the gym facilities, respectively. From the table we get

$$P[\mathbf{E}_\infty^{(\mathrm{use})} \text{ AND } \mathbf{E}_\infty^{(\mathrm{woman})}] = P[\mathbf{E}_\infty^{(\mathrm{use})}] \bullet P[\mathbf{E}_\infty^{(\mathrm{woman})}] =$$

$$= 145/180 \bullet 180/350 = 0.41.$$

Question (b) hypothesizes that the use of gym facilities is influenced by sex. Equation (15.1) and the contingency table yield

$$P[\mathbf{E}_\infty^{(\text{use})} \mid \mathbf{E}_\infty^{(\text{woman})}] = P[\mathbf{E}_\infty^{(\text{use})} \text{ AND } \mathbf{E}_\infty^{(\text{woman})}]/P[\mathbf{E}_\infty^{(\text{woman})}] =$$
$$= (145/350)/(180/350) = 0.80.$$

This high value matches with the actual lifestyle of wealthy women who have much free time which makes it easier to take advantage of the gym facilities.

TLN implies that the frequentist values are real, and the conclusions are facts and not personal credence.

15.5.3 *Single event* (b) – The conditional probability of an individual case is subjective even if it regards a real fact, and available information influences the probability calculus.

Example. Fred is going to roll two dice and is interested in the probability of the sum 9. He considers 36 possible outputs and the following favorable cases: (3,6), (4,5), (5,4), (6,3). Based on this cognition, Fred employs the classical formula and obtains

$$P[\mathbf{E}(\Sigma 9)] = 4/36 = 1/9.$$

In the second play, Fred rolls one die at a time, and means to calculate the probability of the sum 9 when the first roll is 3. Fred is aware that only the combination (3,6) is correct out of the following potential configurations: (3,1), (3,2), (3,3), (3,4), (3,5), (3,6),

$$P[\mathbf{E}^{(\Sigma 9/3)}] = P[\mathbf{E}^{(\Sigma 9)} \mid \mathbf{E}^{(3)}] = 1/6.$$

Fred has developed the calculations in abstract, but means to predict the precise games $\mathbf{E}_1^{(\Sigma 9)}$ and $\mathbf{E}_1^{(\Sigma 9/3)}$ whose probabilities cannot be tested. The numbers which express Fred's opinions are to be rewritten as follows:

$$P[\mathbf{E}_1^{(pr\Sigma 9)}] = 1/9.$$
$$P[\mathbf{E}_1^{(pr\Sigma 9)} \mid \mathbf{E}_1^{(pr3)}] = 1/6.$$

Table 15.1. Conditioning factors and probabilities.

Conditioning factors	Conditional probabilities	
	Frequentist	Subjective
Physical	Y	Y
Cognitive	N.A.	Y

Fred uses the classical formula, which constitutes the 'Prior information' and affects the calculations in both cases.

Subjective probability, conditioned by 'Prior information', qualifies the belief of the individual in the intended fact because P cannot undergo experimental verifications.

15.5.4 As a final brief, physical and cognitive factors can influence prior information of subjective probability; instead, only physical determinants influence the probability of long-term events. Table 15.1 summarizes the structural analysis that bridges the diverse typologies and makes this framework consistent with the frequentist and subjective views.

15.6 Varying Coercive Influence

Conditioning is not a 'yes or no' action but can comprehend various *degrees of coercive influence.*

15.6.1 The event $\mathbf{E}^{(a)}$ can affect more or less profoundly $\mathbf{E}^{(b)}$, it can put more or less pressure on the target and modifies its existence. Conditioning mechanisms reach different degrees of intensity and correspondingly the conditional probability varies in a range. Given $\mathbf{E}^{(a)}$ and $\mathbf{E}^{(b)}$, the probability $P[\mathbf{E}^{(b)} \,|\, \mathbf{E}^{(a)}]$ can assume a spectrum of values.

Example. Tom draws one card from a deck, what is the probability that given a spade ($= A$) he gets the result B?

Conditional probability varies depending on how A can affect B. Suppose B is a face card. There are three face cards (Jack, Queen, and King) in the spades suit and we have $P(A \cap B) = 3/52$,

$P(A) = 13/52$ and (15.1) yields this final result

$$P[\mathbf{E}^{(B)}|\mathbf{E}^{(A)}] = P(B|A) = P(A \cap B)/P(A) = 3/13.$$

Now suppose B is a King instead of the face card, we have

$$P[\mathbf{E}^{(B)} \,|\, \mathbf{E}^{(A)}] = P(B \,|\, A) = 1/13.$$

Suppose B is a number card, we obtain

$$P[\mathbf{E}^{(B)} \,|\, \mathbf{E}^{(A)}] = P(B \,|\, A) = 10/13.$$

In the paradoxical case B is a spade, we get

$$P[\mathbf{E}^{(B)} \,|\, \mathbf{E}^{(A)}] = P(B \,|\, A) = 13/13 = 1.$$

We even consider the limit case where B is no card, therefore, there is no conditioning effect

$$P[\mathbf{E}^{(B)} \,|\, \mathbf{E}^{(A)}] = P(B \,|\, A) = 0/13 = 0.$$

The five cases spell out how the influence of A on B varies between two extremes. The influence is certain when $P[\mathbf{E}^{(B)} \,|\, \mathbf{E}^{(A)}] = 1$ and impossible when $P[\mathbf{E}^{(B)} \,|\, \mathbf{E}^{(A)}] = 0$.

It is evident how an event can influence another event with different degrees of intensity. It is necessary to underscore that there is no general criterion to measure the coercive influence. Probabilists do not have any universal scale of measurement, so this topic deserves further insights.

15.6.2 *Grades from the ontological viewpoint* – $\mathbf{E}^{(a)}$ can physically affect $\mathbf{E}^{(b)}$ with greater or lesser intensity. Let us recall some cases which we read in the literature of engineering, organization science, social science, and so forth.

◼ Suppose the Markovian chain is discrete-time and stationary, the probability of the generic element x_j depends on the previous x_{j-1}

$$P(x_j \,|\, x_{j-1}, x_{j-2}, \ldots, x_1) = P(x_j \,|\, x_{j-1}), \ j = any \ of 1, 2, \ldots, m.$$

This is *first-order dependency*. The conditional probability of the kth order is defined as follows:

$$P(x_j \mid x_{j-1}, x_{j-2}, \ldots, x_1) = P(x_j \mid x_{j-1}, \ldots, x_{j-k}),$$

for $j > k$ and $j =$ any of $1, 2, \ldots$ m,

where $k(k > 1)$ preceding states influence x_j in the chain with memory. In conclusion, the Markovian elements can operate with different coercive influence.

■ A set of random variables are *pairwise independent* when any pair of them are independent. A set of *mutually independent random variables* is pairwise independent, but some pairwise independent collections are not mutually independent.

■ *Hierarchical relations*, which can be more or less coercive, regulate the organizational levels which are typical of companies, institutions, societies, etc. (Kohlberg and Reny, 1997). For the sake of illustration, the head $\mathbf{E}^{(a)}$ can exert a vigorous or feeble pressure over the working collaborator $\mathbf{E}^{(b)}$ and modify the probabilities of the company's outcomes.

Example. An agricultural company has low performances in distributing fresh products D. Mr. X is the *CEO* and the percentage of on-time delivery is about 75%

$$P[D \mid CEO] = 0.75.$$

Mr. Y, the new *CEO*, has increased the salary of the transporters and obtains a higher performance

$$P[D \mid CEO] = 0.98.$$

■ *Local dependencies* make explicit the influence between the members placed at the same level of an organization (Olchi, 1978). For example, the good relations between the colleagues of a plant enhance the work collaboration and the probability of high profits grows.

In conclusion, the conditional probability $P[\mathbf{E}^{(b)} \mid \mathbf{E}^{(a)}]$ can take a spectrum of numerical values because of the varying influence exerted by $\mathbf{E}^{(a)}$ over $\mathbf{E}^{(b)}$.

15.6.3 The *intensity of coercion can change greatly.* On the one hand, the action of $\mathbf{E}^{(a)}$ might be so strong that $\mathbf{E}^{(b)}$ can no longer be extant. On the other hand, $\mathbf{E}^{(a)}$ brings $\mathbf{E}^{(b)}$ to existence as David Hume (2011) fairly writes

> "We may define a cause to be an object, followed by another (...) if the first object had not been, the second never had existed."

In this case, we usually say that $\mathbf{E}^{(a)}$ is *the cause of* $\mathbf{E}^{(b)}$ (Section 14.3). *Causality* can be conceived as the factual and definitive efficacy by which one event establishes the production of another event; causality turns out to be the extreme conditioning situation.

In general, a material event has multiple causes, and the intricate net of factors requires exacting methodologies. Patrick Suppes (1970), Hugh Mellor (1995), Glenn Shafer (1996), and Judea Pearl (2000) have devised criteria, formulas, statistical instruments, rules, and other shrewd methods to help those searching for the causes of an effect. Pearl sets up complex weaponry useful to seek the origin of an uncertain event. He offers a formal and normative description of how rational decisions should be shaped by empirical observations and prior knowledge of one's environment. His account is grounded on causal inference, which he extends beyond probability and statistics. He formalizes *counterfactual reasoning* within a structural-based representation encoding scientific cognition. Counterfactuals are defined as potential yet non-experienced scientific results.

In summary, theoretical works and common experience show how the conditioner can affect an element in various manners and with different degrees of coercion. The scale of the intensity of influence demonstrates that Eqs. (15.1), (15.3), and (15.4) do not exhaust the concept of conditional probability even if they are able to offer aid to problem solvers.

15.6.4 *Grades from the epistemic viewpoint* – For epistemic writers all the probabilities are conditional. Especially, Bayesians have amply developed this topic and I confine myself to citing a few cases just to emphasize the variety of information sources that influence subjective probabilities.

- As first, let us recall the concept of *marginal independence*. Given the variables X and Y, if the knowledge of Y's value does not affect the individual's belief in the value of X, then the random variable X is marginally independent of Y.
- Sometimes, two random variables might not be marginally independent; however, *they become independent* after the involved person observes a third variable. The variable X is conditionally independent of the random variable Y given random variable Z if the knowledge of Y's value does not modify a person's belief in the value of X given the value of Z.
- The *full conditional probability* arises in Bayesian analysis (Cozman, 2013). A Bayesian conditional is generally the distribution of parameters $\theta = (\theta_1, \ldots, \theta_k)$ given the data $y = (y_1, \ldots, y_n)$

$$(\theta_1, \ldots, \theta_k \,|\, y_1, \ldots, y_n). \tag{15.8}$$

When an expert keeps a sample for particular parameters, he conditions not just on the data but also on the current values for every other parameter. For the θ_j^t sample, there is the following conditioning called the *full conditional distribution* of θ_j:

$$\theta_j^t | \theta_1^t, \cdots \theta_{j-1}^t, \theta_{j+1}^t, \ldots, \theta_k^t, y_1, \ldots, y_n. \tag{15.9}$$

Because full conditional distributions are posteriors of conjugates, they are not hard to calculate. Full conditional probability is employed, for instance, to construct a Gibbs sampler, and formulate the distribution of a variable (node) in a *probabilistic graphical model* (PGM) that turns out to be conditioned on the value of all the other variables in the PGM.

- A special position is held by the *Bayesian net* (BN) that probabilists use to handle interfering pieces of information. BN is a directed acyclic graph whose nodes represent observable quantities, latent variables, unknown parameters, or hypotheses (Darwiche, 2009). The edges establish the conditional relationship between the variables. Every node has a conditional probability distribution, that is, the distribution consists of probability values deriving from the states that its parent nodes can assume. BNs are ideal for combining prior information, which often comes in a causal form, and observed data. BNs can be used, even in the case of missing data, to learn the causal relationships and gain an understanding of the various problems as well as predict future events.

Dynamic Bayesian networks prove to be useful for modeling times series and sequences. They extend the concept of standard BNs with time. This graph, in conjunction with Bayesian statistics, provides an efficient approach for avoiding overfitting the data. The numerical semantics of BN offers a quantitative representation of the joint probability distribution in a particular domain of knowledge and expresses it in the form of a numerical indicator.

Despite the adjective '*Bayesian*', BN does not necessarily imply a commitment to Bayesian statistics. Practitioners sometimes estimate the parameters of a BN following frequentist criteria (Pourret *et al.*, 2008).

15.6.5 In conclusion, structural probability theory encompasses different aspects of P, which have caused significant discussion and divisions up to now. The present chapter puts epistemic and ontological conditional probabilities, physical and cognitive factors, and the degrees of coercion inside a unifying framework.

References

Copeland A.N. (1941). Postulates for the theory of probability, *American Journal of Mathematics*, 63, 741–762.

Cozman F. (2013). Independence for full conditional probabilities: Structure, factorization, non-uniqueness, and Bayesian networks, *International Journal of Approximate Reasoning*, 54, 1261–1278.

Darwiche A. (2009). *Modeling and Reasoning with Bayesian Networks* (Cambridge University Press, Cambridge).

de Finetti B. (1974). *Theory of Probability* (John Wiley and Sons, New York).

Hájek A. (2003). What conditional probability could not be, *Synthese*, 137, 273–323.

Hume D. (2011). *An Enquiry Concerning Human Understanding*, E-book (Phoenix Classic).

Kohlberg E. and Reny P.J. (1997). Independence on relative probability spaces and consistent assessments in game trees, *Journal of Economic Theory*, 75(2), 280–313.

Koopman B.O. (1941). The bases of probability, *American Mathematical Society Bulletin*, 2(46), 763–774.

Mellor H. (1995). *The Facts of Causation* (Routledge, London).

Olchi W.G. (1978). The transmission of control through organizational hierarchy, *Academy of Management Journal*, 21(2), 173–192.

Pearl J. (2000). *Causality: Models, Reasoning, and Inference* (Cambridge University Press, Cambridge).

Pourret O., Naïm P., and Marcot B. (eds.) (2008). *Bayesian Networks: A Practical Guide to Applications* (John Wiley and Sons, New York).

Shafer G. (1996). *The Art of Causal Conjecture* (MIT Press, Cambridge, MA).

Suppes P. (1970). *A Probabilistic Theory of Causality* (North-Holland Publishing, Amsterdam).

von Mises R. (1963). *Selected Papers of Richard von Mises*, Vol. 2 (AMS (American Mathematical Society), Providence, RI).

Part 3

Probability and Physics

Chapter 16

The Behaviors of Material Outcomes

Soft disciplines – like sociology, managerial sciences, and political science – usually acquire qualitative data. When they use mathematical models, the experts tend to filter out the numerical results, they lean toward reinterpreting the numbers and often arrive at some compromise. The *ambidexterity metaphor*, recently emphasized as the ability to pursue opposite objectives (Junni *et al.*, 2013), effectively illustrates the elastic thought of managers, sociologists, politicians, humanists, etc. who weigh the numbers using different scales. Naturally, the controversial meanings of probability come to be mitigated for them, and a certain relativism overcomes the frequentism/subjectivism dilemma.

Hard disciplines – say physics, chemistry, mechanics, etc. – behave very differently and do not leave any space for personal appraisals and trade-off. Engineers and technicians expect direct correspondence between the calculated numbers and the real facts.

We are about to investigate the upshots of material events. This chapter addresses physical problems in the classical context, and Chapter 17 examines the quantum context. Various practical cases will corroborate the forecasts of the theorems.

In summary, the third part of this book provides precise answers for the experts in hard disciplines, and experimental evidence will support the theoretical predictions.

16.1 The Statuses of the Outcome

The event and its output are distinct entities; and this separation helps us to investigate the behaviors of the latter from the physical viewpoint (Rocchi and Panella, 2020).

16.1.1 As first, we place the random event on the time scale: $\mathbf{E}^{(e)}$ begins at the start time t_o and winds up at t_e when the process r brings forth the upshot. The following time-intervals are associated with $\mathbf{E}^{(e)}$:

$$T0 = (-\infty, t_o];$$
$$T1 = (t_o, t_e];$$
$$T2 = (t_e, +\infty). \tag{16.1}$$

The outcome is under preparation in T1 and is available in T2. More precisely, the single event $\mathbf{E}_1^{(e)}$ begins at the time t_o and delivers e_1 at t_e. The long-run event $\mathbf{E}_\infty^{(e)}$ starts with the first trial at t_o; it emits the sequence of outcomes e_∞ and finishes with the last trial. If the phenomenon under examination repeats continuously, researchers usually establish the conventional extremes t_o and t_e in accordance with the goals of their investigation.

16.1.2 Let us formulate the statuses of the outcome brought forth by the generic random event.

Definition 16.1. *The result e has the indeterminate status $e^{(I)}$ when one of the following is true:*

$$0 < P(e) < 1; \quad 0 < F(e) < 1. \tag{16.2a}$$

The outcome has the determinate status $e^{(D)}$ when an extreme qualifies it,

$$P(e) = 0; F(e) = 0,$$
$$P(e) = 1; F(e) = 1. \tag{16.2b}$$

16.1.3 The disjoint numerical ranges (16.2a) and (16.2b) entail that they cannot be true together. The two statuses have the property of being mutually exclusive and a middle way is not allowed

Property 16.1.

$$e = e^{(I)} \text{ OR } e^{(D)}. \tag{16.3}$$

The result cannot be determinate and indeterminate at the same time and in the same circumstances.

16.1.4 The time intervals allow the statuses to be accurately related to the time scale. It is apparent that $e^{(I)}$ and $e^{(D)}$ have no material significance during T0 because e is not extant before t_o. The statuses have factual meaning when e is in production during T1 and available in T2, hence we investigate the properties of e in the intervals T1 and T2. We necessarily calculate $P(e)$ in order to establish the status of e during T1, there is no alternative mode. The frequency or empirical probability $F(e)$ is exclusive to T2. For the sake of simplicity, we take $t_o = 0$ from now onward.

16.1.5 Definition 16.1 employs discrete probabilities (Section 10.2). It is worth underlining how the random status (16.2a) is objective and testable even in the single case (Section 12.7).

Theorem 16.1. *Theorem of Initial Conditions (TIC)*
If the event is random, then the outcome is random too in T1

$$0 < P[\mathbf{E}^{(e)}] < 1 \Rightarrow 0 < P(e) < 1, \quad 0 \le t < t_e. \tag{16.4}$$

Proof. Statements (10.8a) and (10.8b) establish equal probability values for the event and its result and prove the theorem.

16.1.6 The ontological–physical perspective of this chapter locates the statuses in precise time intervals and we shall observe carefully whether they persist or change with the passage of time. This type of problems is new in the literature since mathematicians cannot tackle it under the assumption *event = result*.

This chapter is organized as follows:

- Section 16.2 examines the outcome of $\mathbf{E}_\infty^{(e)}$.
- Section 16.3 considers the outcome of $\mathbf{E}_1^{(e)}$.
- Section 16.4 analyzes the outcomes of $\hat{\mathbf{E}}_1^{[N]}$.

16.2 The Outcome of the Long-Term Event

The overall outcome of the repeated event (12.3) is uncertain during T1, and continues in the same status in T2, when $\mathbf{E}_\infty^{(e)}$ no longer runs.

Theorem 16.2. *Theorem of Continuity (TC)*

The outcome e_∞ of the long-term random event $\mathbf{E}_\infty^{(e)}$ keeps the indeterminate status in T1 and T2

$$e_\infty = e_\infty^{(I)}, \quad 0 \leq \quad t. \tag{16.5}$$

Proof. We assume the event is random, and from (16.4), we get

$$0 < P(e_\infty) < 1, \quad 0 \leq t < t_e. \tag{16.6}$$

TLN demonstrates that the relative frequency converges toward probability, hence, the following holds from time t_e forth:

$$0 < F(e_\infty) < 1, \quad t_e \leq t. \tag{16.7}$$

Equations (16.6) with (16.7) demonstrate (16.5) is true.

Let us look into the physical meaning of TC using a case.

Example. Suppose that a coin is flipped 1000 times. The result heads is indeterminate during T1 since $P(H_{1000}) = P(H) = 0.5$. At the end of the experiment, the coin lands H up 473 times. The frequency $F(H_{1000}) = 0.473$, together with $P(H_{1000})$, demonstrates that H_{1000} maintains the indeterminate status in T1 and T2.

16.3 The Outcome of the Single Event

The outcome of the individual elementary event exhibits the strangest conduct. It is uncertain during the process and becomes determinate as soon as $\mathbf{E}_1^{(e)}$ finishes.

Theorem 16.3. *Theorem of Discontinuity (TD)*
The outcome e_1 of the single random event $\mathbf{E}_1^{(e)}$ switches from the indeterminate to the determinate status at the end of the event

$$e_1^{(I)} \to e_1^{(D)}, \quad t = t_e. \tag{16.8}$$

Proof. The event is aleatory and e_1 has the indeterminate status in T1

$$0 < P(e_1) < 1, \quad 0 \leq t < t_e. \tag{16.9}$$

When the event finishes, the allowed values for $F(e_1)$ are zero and one, meaning that e_1 has become certain or impossible in T2. The upshot e_1 goes from the random status to the determinate status in t_e and TD is proved.

Example. The result heads is indeterminate in T1. When the coin lands, $F(H_1)$ equals either 1 or 0, that is to say H_1 switches from being indeterminate to determinate.

Transition (16.8) is not entirely new in the literature. Various ancient philosophers – including Aristotle – noted that ultimately any aleatory phenomenon is either true or false. Randomness is nothing more than a transitory and provisional state, and this observation was enough for thinkers to reject indeterministic reasoning for several centuries, up to the modern era.

The theorem of discontinuity lies at the base of various knotty and paradoxical phenomena which occur in everyday life and in advanced scientific contexts.

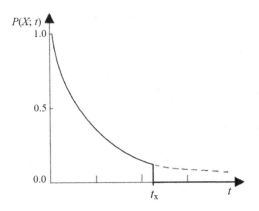

Figure 16.1. Probabilistic status evolution of X with time.

Theorem 16.4. *Theorem of Irreversibility (TI)*
  The status transition (16.8) *is irreversible*

$$P[e_1^{(D)} \to e_1^{(I)}] = 0. \tag{16.10}$$

Proof. The single event finishes with the emission of e_1 and does not regress spontaneously, therefore the transition of e_1, which is strictly bound to the close of $\mathbf{E}_1^{(e)}$, cannot go back.

16.3.1 *Discussing TC and TD* – The theorem of discontinuity proves that the outcome becomes determinate when the aleatory event finishes, that is to say, the outcome becomes either impossible or certain.

Example. Machines and living beings are subject to aging due to multiple and overlapping degradation processes (Rocchi, 2017; Rausand *et al.*, 2020; Johnson and Johnson, 1999). The *reliability function* (also called *survival function*) describes the senescence of system X; specifically, it provides the probability that X will run without failure from time 0 to t_x

$$P(X;t) = e^{-\int_0^t \lambda(t)dt}. \tag{16.11}$$

where $\lambda(t)$ is the *hazard* or *mortality rate function* typical of X. Equation (16.11) expresses the decreasing probability of good functioning with time, which we summarize as follows:

$$0 < P(X;t) < 1, \quad 0 \leq t < t_x. \tag{16.12}$$

When X breaks down at time t_x, the empirical probability of good functioning drops to zero (Figure 16.1)

$$F(X;t) = 0, \quad t_x \leq t. \tag{16.13}$$

The system takes the determinate status, and from (16.12) to (16.13), we get the following result complying with TD:

$$X^{(I)} \to X^{(D)}, \quad t = t_x.$$

Example. Suppose the weight w of babies born in a year varies between $2.1\,\text{kg}$ and $4.7\,\text{kg}$, while the probability density function is approximately constant (Xiong *et al.*, 2007)

$$P(w) = \frac{1}{4.7 - 2.1} = \frac{1}{2.6} = 0.38. \tag{16.14}$$

A baby is born at t_w, the nurse registers his weight w_1 which passes from $P(w_1) = 0.38$ to the value of certitude

$$F(w_1) = 1, \quad t_w \leq t. \tag{16.15}$$

Equations (16.14) and (16.15) demonstrate the shift from the indeterminate to the determinate status that corroborates TD

$$w^{(I)} \to w^{(D)}, \quad t = t_w.$$

16.3.2 The outcome changes status due to the stopping of the event and a simulated stopping also realizes this effect. The single event that halts in a *virtual* manner results in the *virtual* switching of its outcome.

Example. A friend asks Mr. Y: "How old are you?"

Mr. Y provides the following answer: "I am 32 years old."

The survival function (16.11) proves that Mr. Y's status is indeterminate throughout life, but the number 32 fixes the precise state

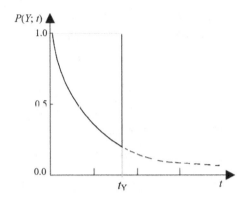

Figure 16.2. Probabilistic status evolution of Y with time.

of life in $t = t_Y$, which is the time of the query. Mr. Y stops his life formally though not in fact (Figure 16.2), and this stop transforms his status from being indeterminate to determinate in agreement with (16.8)

$$Y^{(\mathrm{I})} \to Y^{(\mathrm{D})}, \ t = t_Y. \tag{16.16}$$

16.3.3 The theorem of discontinuity describes an objective effect. Transition (16.8) turns out to be an intrinsic property of the single aleatory process and does not derive from human consciousness.

Example. The tails side of the coin is indeterminate when it is flying in the air. Suppose that the coin lands with tails up at the end of the flight. This result is undisputable even if no observer bends down and looks at it. The status transition is a process which does not derive from human awareness.

The theorem of discontinuity predicts that the probability status changes regardless of human will. TD ignores the observer who can intervene anyway in the physical reality. For instance, an agent causes the halt of $\mathbf{E}_1^{(e)}$ and makes the change (16.8) due to operational reasons. May be an individual executes a risky operation that involuntarily brings $\mathbf{E}_1^{(e)}$ to finish.

When an event ends in the real world, the individual recognizes that e_1 has become certain or impossible. This incontrovertible change improves human knowledge and does not come from human consciousness.

Perhaps the reader finds the present comments rather pedantic; it is true, but the comments are necessary because of the ontological and epistemic perspectives, which often mix up, especially in the quantum field.

16.4 The Outcomes of the Quasi-Certain Event

The theorem of discontinuity can be deemed as an essential theoretical result. This topic needs more extensive inquiries and we look into the quasi-certain structure which provides the complete account of aleatory events.

16.4.1 The theorem of complete structure proves that $\hat{\mathbf{E}}^{[N)}$ certainly brings forth the outcome e

$$\hat{\mathbf{E}}^{[N)} = [\mathbf{E}^{(e1)} \text{ OR } \mathbf{E}^{(e2)} \text{ OR } \ldots \text{OR } \mathbf{E}^{(eN)}] = (i, r - e). \qquad (9.9)$$

which consists of N alternative possible outcomes

$$e = (e_1 \text{ OR } e_2 \text{ OR } e_3 \text{ OR } \ldots \text{ OR } e_N). \qquad (9.11)$$

The outcomes $e_1, e_2, e_3 \ldots e_N$ form the aggregate (9.11) on paper, whereas they can make a physical group in the world or not. They may or may not coexist during T1 and the *coexistence* turns out to be a very important feature because the event $\hat{\mathbf{E}}_1^{[N)}$ prepares the outcomes using different techniques. For example, it makes ready the outcomes step by step, or creates an outcome by assembling the pre-packaged components and so forth. An important kind of quasi-certain event handles $e_{11}, e_{12}, e_{13}, \ldots, e_{1N}$ all together during the interval T1. This situation is typical of the games of chance whose outcomes are potentially ready since the beginning of the game. *All results are factually available during* T1 and overlap in some sense. The following set, formalized by the brackets $\langle \rangle$, specifies that the results are *superposing*:

Definition 16.2. *If the N potential results of $\hat{\mathbf{E}}_1^{[N]}$ are ready during T1, they make a set of superposed random results*

$$\langle e_{11}^{(I)}, e_{12}^{(I)}, e_{13}^{(I)}, \ldots, e_{1N}^{(I)} \rangle, \quad t < t_e. \qquad (16.17)$$

Example. The roulette wheel exhibits the numbers from 0 to 36 and all of them are available for extraction during the rotation of the wheel, namely the numbers superpose during T1

$$\langle 0_1^{(I)}, 1_1^{(I)}, 2_1^{(I)}, \cdots 35_1^{(I)}, 36_1^{(I)} \rangle.$$

16.4.2 The next theorem demonstrates that the ensemble of overlapping upshots decreases in size when the experiment is over.

Theorem 16.5. *Theorem of Reduction (TR)*
  If $\hat{\mathbf{E}}_1^{[N]}$ has N superposed outcomes, they reduce to just one determinate outcome at the end of T1

$$\langle e_{11}^{(I)}, \ e_{12}^{(I)}, e_{13}^{(I)}, \cdots e_{1j}^{(I)} \cdots, e_{1N}^{(I)} \rangle \to \{ e_{h1}^{(D)} \},$$

$$\text{where } j = \text{any of } 1, 2, \ldots N; \ t = t_e. \qquad (16.18)$$

Proof. All possible outcomes are simultaneously available during T1, and we apply the theorem of discontinuity to each one. If $\hat{\mathbf{E}}_1^{[N]}$ brings forth $e_{1j}(j = \text{any of } 1, 2, \ldots, N)$ at time t_e, then e_{1j} becomes *certain*. The remaining potential outcomes also conform to TD since they become *impossible* due to the normalization rule. All potential outcomes switch from the indeterminate to the determinate status, and the following equations make explicit (16.18):

$$[0 < P(e_{1j}) < 1] \to [F(e_{1j}) = 1],$$

$$j = \text{any of } 1, 2, \ldots, N. \qquad (16.19a)$$

$$[0 < P(e_{1k}) < 1] \to [F(e_{1k}) = 0],$$

$$k = 1, 2, \ldots (j-1), (j+1), \ldots, N; \quad t = t_e. \qquad (16.19b)$$

Example. An organization holds a lottery and sells N tickets. All the tickets are candidates for the victory

$$\langle c_1^{(I)}, c_2^{(I)}, c_3^{(I)}, \ldots, c_N^{(I)} \rangle, \quad t < t_c.$$

with the following chance to win:

$$0 < [P(e_{1j}) = 1/N] < 1, \quad j = \text{any of } 1, 2, \ldots, N.$$

When the time limit of *the lottery* expires, the *extraction* takes place, and the ensemble of potential winners reduces in size: one wins, the remaining tickets lose and cease to exist.

16.4.3 The cluster of upshots decreases in size only when they are simultaneously ready in T1; if not, the upshots do not make a real cluster and consequently *the cluster cannot downsize*. If the event outputs only one result at time, this feature does not imply that the hypothesis of superposition is true. The following case helps the reader to grasp the peculiarity of TR.

Example. Mr. Z is a person who, generically speaking, takes on the *living status* $Z_L^{(I)}$ and the *dying status* $Z_D^{(I)}$, which are mutually exclusive

$$Z = (Z_L^{(I)} \text{ OR } Z_D^{(I)}). \tag{16.20}$$

The survival function (16.11) demonstrates that Z_L and Z_D are aleatory states

$$0 < P(Z_L; t) < 1,$$

$$0 < \{P(Z_D; t) = [1 - P(Z_L; t)]\} > 1, \quad 0 < t < t_Z. \tag{16.21}$$

As soon as Mr. Z dies at t_Z, we obtain

$$F(Z_L; t) = 0,$$

$$F(Z_D; t) = 1, \quad t_Z \le t. \tag{16.22}$$

The double transition from (16.21) to (16.22) seems to comply with the theorem of reduction

$$[0 < P(Z_L; t) < 1], \rightarrow [F(Z_L; t) = 0];$$

$$[0 < P(Z_D; t) < 1], \rightarrow [F(Z_D; t) = 1]. \tag{16.23}$$

Instead, the states $Z_L^{(I)}$ and $Z_D^{(I)}$ do not superpose, Mr. Z is not half alive and half dead during his lifespan and the theorem of reductions does not hold. The states $Z_L^{(I)}$ and $Z_D^{(I)}$ are never simultaneous and cannot reduce in number, (16.23) simply proves they alternate at the instant t_Z.

This remark should be extended to the well-known 'Schrödinger's cat paradox' because a cat's life does not conform to the superposition hypothesis (Schrödinger, 1980). This long-debated argument does not have ground here and sounds like a baseless metaphor from the present viewpoint.

16.4.4 *Again about testability* – Chapter 12 analyzes the possibilities of testing probability in the world. The present chapter goes further in the same direction and investigates the conduct of physical random results, so that the theorems of Chapters 12 and 16 make two groups:

- The theorems of large numbers and continuity ensure that testing can corroborate the probability of the outcome of the long-term event.
- Three theorems revolve around the single outcome and describe it from different stances which complete one another. The theorem of a single number proves that $P[\mathbf{E}_1^{(e)}]$ cannot be tested, the theorem of discontinuity demonstrates that when the single trial is over, the outcome e_1, till then uncertain, suddenly becomes determined. The theorem of reduction proves that only one potential outcome becomes certain while the residual outcomes vanish.

TLN and TSN regard the exact numbers $P(e_\infty)$ and $P(e_1)$. TC, TD and TR regard generic decimals that are testable no matter the number of events

$$0 < P(e_\infty) < 1,$$
$$0 < P(e_1) < 1.$$

In fact, Section 12.7 has shown how the haphazardness of the single and long-term events is objective and can be verified using various techniques.

16.4.5 Some scholars place confidence in probability calculus as a comprehensive mathematical support. The fifteen theorems proved in this book point out topics and phenomena which have so far remained in the shadow.

References

Johnson R.C. and Johnson N.L. (1999). *Survival Models and Data Analysis* (John Wiley, New York).

Junni P., Sarala R.M., Taras V., and Tarba S.Y. (2013). Organizational ambidexterity and performance: A meta-analysis, *The Academy of Management Perspectives*, 27(4), 299–312.

Rausand M., Barros A., and Hoyland A. (2020). *System Reliability Theory: Models, Statistical Methods and Application*, 3rd edition (John Wiley, New York).

Rocchi P. (2017). *Reliability is a New Science: Gnedenko was Right* (Springer, Berlin; New York).

Rocchi P. and Panella O. (2020). Some probability effects in the classical context, *Open Physics*, 18(1), 512–516.

Schrödinger E. (1980). A translation of Schrödinger's 'Cat Paradox', *Proceedings of American Philosophical Society*, 124, 323–338, Available at http://hermes.ffn.ub.es/luisnavarro/nuevo_maletin/Schrodinger_1935_cat.pdf.

Xiong X., Wightkin J., Magnus J.H. Jeanette H. Magnus, Pridjian G., Acuna J.M., and Buekenset P. (2007). Birth weight and infant growth: Optimal infant weight gain versus optimal infant weight, *Maternal and Child Health Journal*, 11, 57–63.

Chapter 17

An Essay on Quantum Mechanics

The current chapter is arranged in four parts:

- Unit I illustrates the contents and aims of this study.
- Unit II applies the theorems of the structural theory to quantum mechanics.
- Unit III deals with the quantum experiments that corroborate the theoretical forecasts.
- Unit IV provides the conclusions.

UNIT I – CONTENTS AND AIMS OF THE PRESENT ESSAY

17.1 Lively Debates

Since its birth, quantum physics has raised knotty questions, meaning that existing theories seem incapable of explaining some experimental results. While some advanced problems have been brilliantly solved, base issues continue to be the subjects of lively debates within the community of scientists and philosophers (Myrvold, 2016; Squires, 1994). This chapter briefly recalls the questions, which it aims to address.

17.1.1 *What are particles and waves?* – Scientists agree that quanta are discrete portions of energy and occasionally of matter, which assume two states. Particles and waves provide rational descriptions fitting different experimental situations, but the current interpretations of quantum mechanics (QM) do not provide satisfactory answers since neither particles nor waves provide an exhaustive explanation of physical reality:

> "It seems as though we must use sometimes the one theory and sometimes the other, while at times we may use either. (...) We have two contradictory pictures of reality; separately neither of them fully explains the phenomena of light, but together they do." (Einstein and Infeld, 1938)

Certain authors such as Bohm, Hiley, and de Broglie sustain both the states; some view only waves, such as Mead, Horodecki and Everett; and other authors regard only particles such as Duane, Du and Landé.

Several theorists try to go beyond this deadlock by means of very abstract constructions. They assume that a wavefunction is a mathematical object corresponding to an element of a certain Hilbert space of infinite dimension that groups the possible states of the system. Therefore, quantum objects are not specifiable at the physical level; they are neither waves nor particles; they are a strange combination of both (Rae, 2004).

17.1.2 *Why does the wave collapse? What is measurement?* –

When an operator tries to meter the wavefunction, it collapses, and the operator always finds a particle located in an unpredictable position (Boughn and Reginatto, 2013).

Several authors are inclined to believe that the quantum wave, initially in a superposition of many eigenstates, reduces to a single eigenstate due to its interaction with some measuring action. These authors assume that the wavefunction is a linear combination of the eigenstates of an observable where the coefficients c_1, c_2, c_3, ... are the probability amplitudes corresponding to each eigenstate

$$|\Psi\rangle = c_1|\psi_1\rangle + c_2|\psi_2\rangle + \cdots + c_j|\psi_j\rangle + \cdots + c_r|\psi_r\rangle.$$

They conclude that $|\Psi\rangle$ downsizes when it collapses

$$c_1|\psi_1\rangle + c_2|\psi_2\rangle + \cdots + c_j|\psi_j\rangle + \cdots + c_r|\psi_r\rangle \to |\psi_h\rangle,$$
$$j = \text{any of } 1, \ldots r.$$

Why the collapse takes place instantaneously, how it occurs, and what its physical cause is, are questions that remain unanswered. Consequently, some – the Copenhagen interpretation and others – have decided to formalize the collapse as a postulate, namely, as a non-provable statement (Cushing, 1994). Other authors – e.g., Born, Birkhoff, Watanabe and Everett – even negate the collapse.

The measurement equipment and the measured system seem to form an integrated and rather mysterious whole which some schools formalize together with the observer. A circle of thinkers, including Niels Bohr (1961), answers the questions in radical terms and outlines an abstract vision deprived of any causal principle.

17.1.3 *Is quantum physics real science?* – Physicists adopt a mode of research in which any hypothesis must be controlled, carefully documented, and replicable with experiments; hence, the measurement problem calls into question the scientific status of QM. Two prominent accounts dating back to the 1920s express opposing conclusions about the scientific essence of quantum physics (Colbeck and Renner, 2012):

A] In one view, the wavefunction does not represent reality but represents an observer's state of knowledge about some underlying reality. Scientists inspired by the Bayesian theory judge the collapse not as an operation but as a source of knowledge about "the potential consequences of our experimental interventions into nature" (Peierls, 1991). "Consciousness causes collapse" is the speculative conclusion of those conceiving the observation made by man as responsible for the transformation of the quantum system

> "I believe that one can formulate the emergence of the classical 'path' of a particle succinctly as follows: the 'path' comes into being only because we observe it." (Heisenberg, 1927)

The human mind and quantum objects would be entangled, and one cannot be considered apart from the other (Fuchs and Peres, 2000). Only the mathematical formalism is meaningful, and any attempt at providing a material interpretation is considered nonsense or a metaphysical way of handling the problems. These thinkers arrive at the supposition that the monitoring system does not serve to check a quantum variable; it *creates* a particular value for that variable. This *'measurement = creation principle'* constitutes the ultimate cultural repercussion of the dramatization of the measurement problem (Boniolo, 2000).

B] From the alternative stance, the wave function corresponds to an element of reality that objectively exists whether or not an observer is controlling it. Several eminent authors share this position. For instance, Einstein declares that it is the purpose of physics to provide an objective description of reality and he puts forward a *hidden-variable* interpretation of QM to go beyond the intellectual standstill which the present unresolved dilemmas have created. Roger Penrose develops a realistic cognition of QM related to general relativity. Various constructions such as the de Broglie–Bohm theory, the transactional interpretation, and the objective collapse theories share viewpoints close to 'realism'.

17.1.4 *Is quantum physics a limit case?* – In point of logic, there should be continuity or symmetry between QM and classical mechanics (CM). Niels Bohr (1949) posited the *correspondence principle,* which remains a controversial doctrine up to this day. Several ideas have been proposed to show CM as a limiting case of QM (Landsman, 2007), such as the following:

- The behavior of systems described by QM should reproduce CM within the limit of large numbers (Gregg, 2014). There should be an asymptotic statistical agreement between the classical frequency and the quantum frequency. However, this correspondence is legitimated only if the system is in certain 'classical' states and monitored with 'classical' observables.
- The Planck constant h, together with the mass and other parameters, determines the characteristics and the temporal scale of the quantum motion. We could hypothesize that h becomes small when the wavefunction describes the probability density of a particle that is smeared out around a classical trajectory.

The smaller the h, the smaller the scale where a quantum phenomenon manifests itself, and in turn the particle motion is more localized along the classical trajectory.

- *Coherent histories* also known as '*consistent histories*', seek to generalize the *Copenhagen interpretation* and integrate with *quantum decoherence* which implies that irreversible macroscopic phenomena – hence, all classical experimental controls – make histories automatically consistent. It should become natural to recover classical reasoning and 'common sense' applied to the outcomes of the quantum measurements, but the application of the above-mentioned criteria turns out to be problematic (Zurek, 2003).

In principle, the quantum–classical compatibility sounds somewhat clear, however, the various answers that have been put forward are anything but satisfactory. Despite noteworthy research efforts, no explanation regarding the correspondence principle is considered definitive (Ghose, 2002).

17.2 A Question of Method

Engineers have built up nuclear plants, medical equipment, and other brilliant solutions; at the same time, theorists meet heavy obstacles in explaining the most elementary parts of energy and matter. Satisfactory illustrations give the impression of moving away instead of coming close:

> "Surely after 62 years we should have an exact formulation of some serious part of quantum mechanics. By 'exact' I do not mean of course 'exactly true'. I only mean that the theory should be fully formulated in mathematical terms, with nothing left to the discretion of the theoretical physicist ...". (Bell, 1990)

I personally believe that the "exact formulation of some serious part of quantum mechanics" wholly depends on the research method.

Let us analyze three methodological issues.

17.2.1 *The head of the knot* – Physicists who pioneered QM addressed the questions one by one as they came up. The history of science shows how unexpected results, curiosities, fortuitous findings, innovative experiments, etc. drove the progress of QM in the first decades. Nowadays, the base issues create an inextricable Gordian knot (Sections 17.1.1 to 17.1.4) and we cannot continue with the same exploring style. We must begin with the head of the knot that regards the tiny objects investigated by quantum scientists. They constitute the first riddle to tackle; indeed, pioneers began by asking:

What is quantum physics dealing with?

In 1900, Max Planck, assuming that energy could be released in discrete packets, solved the blackbody problem. Shortly thereafter, Einstein, Millikan, Thomson, Compton, Rayleigh, and others made the scientific community fully aware of indivisible elements characterized by multiples of a basis value. This book calls 'quantum' a whole photon (or electron, proton, etc.) or a fraction thereof that conforms to the following principle:

[1] *The quantum ξ is a discrete quantity of energy and possibly matter.*

A black body absorbs and emits all radiation frequencies, and Einstein (1909) investigated the statistical fluctuations of energy whose variance adds two values

$$\sigma^2 = \bar{n} + \bar{n}^2. \qquad (17.1)$$

The term $\bar{n}$ refers to the fluctuations of a group of independent particles, and $\bar{n}^2$ corresponds to the fluctuations of classical waves in a cavity, that is to say, the radiation of the black body exhibits both particle and wave aspects. Numerous subsequent experiments have made quantum duality evident, which can be expressed as follows:

[2] *The quantum ξ behaves either as a wave or as a particle.*

$$\xi = (\textit{Wave} \text{ OR } \textit{Particle}).$$

In summary, ξ has three characteristics: the first is unique and the other two are mutually exclusive (Figure 17.1). Physicists manage the quantization principle by using integral and half-integral values that identify the state of a physical system, instead the weird wave–particle states of ξ constitute an ongoing conundrum that should be resolved by precise definitions.

Figure 17.1. Logical map of the quantum determination.

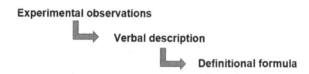

Figure 17.2. Procedure for a definitional equation.

17.2.2 *Definitional equations* – When physicists explore a new sector, they determine the elements pertaining to that sector in formal terms. Basically, scientists follow this procedure: they begin with experimental observations; they identify and describe the elements under scrutiny; and finally translate the shared understanding into the mathematical language (Figure 17.2).

The *definitional formula* (DF) of the physical quantity x has the special capability of describing the intrinsic nature of x. The *computational formula* (CF) describes an aspect of x or provides further insights into x, but does not state the basic qualities of x. DFs and CFs provide effective aid to scientists in applied problems, but only DF explicates the essence of x and ensures our understanding of x. For example, Ohm's law and the electric power allow the calculation of the electric current of countless circuits

$$I = V/R.$$
$$I = W/V.$$

However, only the DF makes explicit the nature of I, which is the electric charge rate flowing through a conductor

$$I = \frac{dQ}{dt}.$$

DFs turn out to be indispensable for understanding the objects under examination and it is natural to wonder:

What are the DFs of the quantum particle and wave?

Various ideas are circulating. For some, the particles are vibrating strings; for others, they are an irreducible representation of the Poincaré group. Within the framework of the first quantum physics, the *particle is a wavefunction that propagates in space*. Let us delve deep into this topic amply shared by the scientific community.

The wavefunction furnishes the mathematical solution to the Schrödinger equation, which is symmetrical to the classical criterion of energy conservation. The general form of the equation includes the Hamiltonian $\hat{H}$ that has the property of being Hermitian. The *time-independent form* of the equation describes standing waves, while the *time-dependent equation* describes progressive waves

$$E|\Psi\rangle = \hat{H}|\Psi\rangle. \tag{17.2a}$$

$$i\hbar\frac{\partial}{\partial t}|\Psi(t)\rangle = \hat{H}|\Psi(t)\rangle. \tag{17.2b}$$

The *position wavefunction* $\Psi(x,t)$ solves the Schrödinger equation in spatial space, and the *momentum wavefunction* $\Phi(p,t)$ supplies the solution in momentum space

$$i\hbar\frac{\partial}{\partial t}\Psi(x,t) = -\frac{\hbar^2}{2m}\frac{\partial^2}{\partial x^2}\Psi(x,t) + V(x)\Psi(x,t). \tag{17.3a}$$

$$i\hbar\frac{\partial}{\partial t}\Phi(p,t) = \frac{p^2}{2m}\frac{\partial^2}{\partial x^2}\Phi(p,t) + \tilde{V}(p)\Phi(p,t). \tag{17.3b}$$

They have mutual relationships

$$\Psi(x,t) = \int_{-\infty}^{+\infty} e^{ixp/\hbar}\,\Phi(p,t)\,dp.$$

$$\Phi(p,t) = \frac{1}{2\pi\hbar}\int_{-\infty}^{+\infty} e^{-ixp/\hbar}\Psi(x,t)\,dx.$$

The wavefunctions can qualify bound particles, such as electrons in the shells of an atom, or free elements such as α and β particles. Quanta with internal properties – such as the spin of a bound electron or the angular momentum of a photon – are described by wavefunctions with several components. Depending on the transformation behavior of the functions in Lorentz transformations, a distinction is made in the relativistic quantum field theory among scalar, tensorial, and spinor wavefunctions.

These concise remarks should be enough to conclude that the wavefunctions furnish the mathematical solutions to the Schrödinger equation and other problems but do not stem from experimental descriptions and therefore do not have the requisite typical of DFs. The *wavefunctions are computational formulas and are not definitional.*

This conclusion matches with the debates of quantum physicists who are notoriously divided when it comes to the correct interpretation of $\Psi(x, t)$ and $\Phi(p, t)$. What the wavefunctions represent, whether they really exist, if they are purely formal representations, etc., are major arguments of QM. If the wavefunctions were DFs, these doubts would not be raised.

Those who take the wavefunctions as DFs tend to draw generic conclusions and make contradictory claims. For example, they present $\Psi(x, t)$ as a *particular condition of the particle* and negate [2] that posits the wave as a *particular condition of the quantum.* Writers often state that "the particles exhibit both wave and particle properties" while the correct sentence should be: "quanta exhibit both wave and particle properties". Experts also talk about "the frequency and the wavelength of the particle". which is an evident 'contradiction in terms'; the particle as such cannot have any frequency and wavelength, which are typical features of the wave.

In conclusion, $\Psi(x, t)$ and $\Phi(p, t)$ are not DFs, and in a rather paradoxical and provocative manner we could say that "QM does not know the objects that it is studying". *The definitional formulas of quantum physics are missing* and what is worse, there is another relevant methodological prerequisite.

17.2.3 *Mathematics first* – Chapter 5 recalls how mathematicians and physicists cooperate in a precise way. The latter can work out a theoretical solution if and only if mathematicians devise the necessary formal instruments in advance. If mathematics does not hit the target, physics cannot go ahead, and this rule does not leave any room for personal discretion or trade-offs.

Unfortunately, some physicists believe that probability calculus functions perfectly and deem the divisions among the probabilists are no more than philosophical quarrels. Those physicists fail to heed the negative repercussions of the fragmentary and incomplete probability theories on QM. They find it difficult to grasp the present cultural situation, which can be compared to the birth of CM.

Nobody can handle traditional mechanical problems without the aid of derivatives and integrals, but infinitesimal analysis was not invented until the middle of the 17th century when Newton, and independently Leibniz, discovered the 'calculus of infinitesimals'. Ten years later the Englishman was able to publish the laws of mechanics with the help of the new mathematical instruments.

The history of physics helps us to be more specific. The *vectorial theory* underpins the definitions of motion, velocity, acceleration, and force, and explains how these parameters are not scalar but are characterized by magnitude, direction, and sense. At the same time, infinitesimal calculus provides powerful CFs in CM (Whiteside, 1970). Currently, the probability theory is expected to underpin the DFs of QM, while linear operators, spectral theory, complex numbers, Hilbert spaces, etc. solve advanced issues.

The groundbreakers of CM were somewhat familiar with the composition of vectors and the parallelogram rule that paved the way toward DFs, whereas infinitesimal calculus was not developed until the late 17th century. Nowadays, it may be said that the intellectual situation has turned upside down because quantum theorists can develop important CFs but are unable to state the DFs using probability. The fragmentary theories prevent them from explicating the nature of quantum waves and particles (Table 17.1) while they are familiar with sophisticated mathematical weaponry and calculate complex quantum systems. Great scientists – Einstein, Dirac, Bohm, Schrödinger, and others, who came up with ingenious solutions – could not clarify the fundamentals of QM because they lacked the exhaustive probability theory. They, too, could not take shortcuts that are not there.

Table 17.1. How mathematics sustained the pioneers of classical and quantum mechanics.

		For Pioneers		For Pioneers
CM	Vector theory	*Ready*	Infinitesimal calculus, etc.	*Unready*
QM	Probability theory	*Unready*	Linear algebra, differential calculus, functional analysis, etc.	*Ready*

17.2.4 In conclusion, the methodological annotations 17.2.1, 17.2.2, and 17.2.3 can be summarized as follows:

X) The wavefunctions are CFs and not DFs.
Y) Probability should underpin the quantum DFs.
Z) The partial theories of probability do not support quantum DFs.

The ample theory developed in the previous pages offers a chance to surmount the dramatic obstacle **Z)** and can be used to undertake the description of quanta (Rocchi and Panella, 2021, 2022). Obviously, this essay is a purely probabilistic inquiry which does not calculate physical parameters such as energies, speeds, momenta, etc. It will also overlook advanced problems which lie beyond the scope of the present work.

UNIT II – DEFINITIONAL FORMULAS IN QM

17.3 Descriptions of the Quanta with Probability

A definitional formula is obtained from experimental observations, and we necessarily follow this procedure (Section 17.2.2).

17.3.1 The majority of scientists agree on these points:

(i) Werner Heisenberg has formulated the *principle of uncertainty*, which is routinely observed in many experiments. He states that the conjugates A and B are not simultaneously known to an arbitrary precision.

$$\Delta A \cdot \Delta B \geq \frac{\hbar}{2}. \tag{17.4}$$

This objective inability to make precise measurements implies that the quantic entities have an intrinsic indeterministic nature.

(ii) The quantum ξ is a discrete portion of energy and at times of matter, which takes up the domain Λ of the Euclidean space Σ. When ξ is a particle, it occupies a spot that has almost zero volume V_P; when ξ is a wave, it has the quality of being pervasive and the volume of Λ can be, in principle, infinite

$$V_P \approx 0; \tag{17.5a}$$

$$V_W \neq 0. \tag{17.5b}$$

The *quantization principle* holds that ξ cannot be subdivided into smaller parts, hence the spatial shapes of (17.5a) and (17.5b) must be necessarily expressed in probabilistic terms.

17.3.2 Both observations (i) and (ii) yield that the definitions of quanta are to be formulated by means of P, and immediately we face an obstacle since the *structural probability* P is not fully in accord with the *quantum probability* Pr (Gudder, 1988; Kümmerer and Maassen, 1998; Khrennikov, 2016), which owns the following unconventional attributes:

   I. Pr does not have commutative properties.
  II. Pr is calculated on the basis of complex numbers.
 III. Pr can have negative values.
 IV. Pr admits non-local relations.

Structural probability and quantum probability do not coincide, and one may doubt the reliability of P in the quantum context.

In reply, I note the following shared properties:

a Both P and Pr qualify the existence of events and outcomes.
b Both P and Pr use integers to qualify the impossible and the certain events; they use decimals for random cases.

In this chapter, the structural and the quantum probabilities deal with the occurrence of physical elements and, in doing so, they are confined to *one*, *zero*, and the *generic decimal*. The three values have parallel meanings for P and Pr that prove to be compatible here.

17.3.3 *Quantum statuses* – In conformity with (i) and (ii), and remarks $\mathfrak{a}$ and $\mathfrak{b}$, it is reasonable to introduce $P[\xi \in (x, y, z)]$, that is the probability of finding ξ in the point (x, y, z) of Σ at a generic time.

The preliminary step imports expressions (16.2a) and (16.2b) into QM:

Definition 17.1. *The quantum ξ has the indeterminate status $\xi^{(I)}$ when one or both of the followings are true*

$$0 < P[\xi \in (x, y, z)] < 1; \quad 0 < F[\xi \in (x, y, z)] < 1. \tag{17.6}$$

ξ has the determinate status $\xi^{(D)}$ when the extremes qualify it

$$P[\xi \in (x, y, z)] = 0; \quad F[\xi \in (x, y, z)] = 0,$$
$$P[\xi \in (x, y, z)] = 1; \quad [\xi \in (x, y, z)] = 1. \tag{17.7}$$

The reader can note how ξ is conceived as a result assessed by probability calculus. For the present theory ξ is an observable entity not precisely determined in terms of energy and matter, because this theory follows a purely probabilistic approach and ignores material indices such as speed, mass, etc. It also disregards the mathematical instruments and models necessary to handle such indices.

17.3.4 *DFs of quantum particles and waves* – Equation (17.5a) implies that the particle is *localized*, it occupies a precise position, namely, it has the determinate status in spatial terms. The wave is *not localized* and embodies the indeterminate status of ξ.

Definition 17.2.

$$Particle = \xi^{(D)}. \tag{17.8}$$

$$Wave = \xi^{(I)}. \tag{17.9}$$

These definitions conform with the normalization rule, hence (17.8) generates the *spatial probability distribution* $P_{\xi D}$ with a unity in the point (x', y', z') and zero in the remaining points of space:

$$P_{\xi D} = \begin{cases} P[\xi \in (x', y', z')] = 1, \\ P[\xi \in (x, y, z)] = 0, \quad x, y, z \neq x', y', z'; \; x, y, z \in (-\infty, +\infty). \end{cases}$$

$$(17.10)$$

Delta function (17.10) is the universal distribution for particles. Instead, the general distribution function for waves cannot be fixed since the shape of the wave depends on specific physical constraints. We can make explicit (17.9) using $P_{\xi I}$ where (x, y, z) is the generic point of the domain Λ:

$$0 \leq \{P_{\xi I} = P[\xi \in (x, y, z)]\} < 1. \qquad (17.11)$$

Sections 17.3.6 and 17.6.7 will provide further insights about this topic.

The numerical values of (17.6) and (17.7) imply that ξ can be either a particle or a wave, and this is consistent with principle **[2]** that establishes two alternative states of the quantum.

Property 17.1

$$\xi = (\xi^{(I)} \text{ OR } \xi^{(D)}). \qquad (17.12)$$

This property denies that: "the wave is a special condition of the particle", instead, the wave and the particle are *special conditions of* ξ.

17.3.5 Definitions (17.8) and (17.9) express ideas that are shared since long. Let us cast an eye on the literature which reverberates the present perspective.

Particle physics is a branch of physics that investigates the elementary constituents of matter (and antimatter) and radiation, and the interactions between them (Dodd and Gripaios, 2020). Modern research focuses on subatomic particles and a wide range of exotic corpuscles that, in principle, are consistent with the model (17.10). The *Standard Model* (SM) is probably the best theory for describing the basic building blocks of the universe. It explains how quarks make up protons, neutrons, and other particles which in turn form atoms. SM explains three of the four fundamental forces that govern the universe: electromagnetism, the strong force, and the weak force (Nachtmann, 2012).

Fundamental particles possess properties such as electric charge, spin, mass, magnetism, and an assortment of characteristics that distinguish one from another. Despite great differences, all of them share the quality of being point-like as $\xi^{(\mathrm{D})}$.

Example. Suppose the electron ξ with charge q and speed v runs perpendicular to the uniform magnetic field B. The magnetic force $F_B = qvB$ supplies the centripetal force $F_C = mv^2/\rho$

$$qvB = \frac{mv^2}{\rho}.$$

We obtain the radius of the resulting circular path of the electron

$$\rho = \frac{mv^2}{qvB} = \frac{p}{qB}.$$

The precise orbit makes explicit the punctiform shape of the electron in conformity with $\xi^{(\mathrm{D})}$.

Quantum Electrodynamics (QED) is the relativistic field theory dealing with two fundamental objects: light and matter (Scharf, 2014). QED is an Abelian gauge theory with the symmetry group which formally describes the interactions of light with matter and the interactions of charged quanta with one another. The fields are functions of space-time coordinates, so except for very exceptional circumstances, the excitations of the electromagnetic field or the electron field are not spatially localized, they are everywhere in agreement with the concept $\xi^{(\mathrm{I})}$. Depending on how a particular interaction is arranged, an excitation may have a well-defined position, but the ordinary state of these excitations is that they are spread out in space.

In the literature we find strands of theoretical studies focusing on quantum waves which exhibit spreading and fluctuating shapes as $\xi^{(\mathrm{I})}$.

Example. Suppose a quantum wave is moving horizontally within an infinitely deep well from which it cannot escape. For the normalized stationary states of energy $(n = 1, 2, 3, \ldots)$, we get

$$\Psi_n(x) = \sqrt{\frac{2}{a}} \sin\left(\frac{n\pi}{a}x\right)$$

whose oscillations specify the distributed shape of $\xi^{(\mathrm{I})}$.

17.3.6 Coherent set of equations – Definitions 17.2 make explicit the probabilistic nature of quantic entities that are directly related to the following equations.

■ The Heisenberg principle (17.4) specified for the position x and the momentum is the following:

$$\Delta x \cdot \Delta p \geq \frac{\hbar}{2}. \tag{17.13}$$

Louis de Broglie (1950) introduced a theory of mechanics according to which the quantum ξ behaves like a wave whose wavelength is inversely proportional to the momentum p

$$\lambda_B = \frac{h}{p}. \tag{17.14}$$

This pair of equations implies that if the quantum ξ has the precise position x, then it is pulse-like in harmony with $\xi^{(\mathrm{D})}$ and has an improper momentum. Instead, when ξ has the accurate value of p, we get the precise wavelength λ_B from (17.14); ξ spreads throughout space and its position is ill-defined in agreement with $\xi^{(\mathrm{I})}$.

■ We can expand a *position eigenfunction* based on *momentum eigenfunctions* and vice versa. For the particle $\xi^{(\mathrm{D})}$ having the definite position x' in 1D space, its normalized wavefunction can be expanded on the momentum eigen basis, which, in principle, can be done using the Fourier transform

$$\Psi_{x'}(x) = \delta(x - x'). \tag{17.15}$$

However, we would need an infinite number of wave terms with definite values of the momentum to express the perfectly localized delta function

$$\Psi_{x'}(x) = \delta(x - x') = \int_{-\infty}^{+\infty} \frac{dp}{2\pi\hbar} \, e^{\frac{i}{\hbar}p(x-x')}. \qquad (17.16)$$

On the other hand, for the wave $\xi^{(\mathrm{I})}$ with definite momentum, we get

$$\Phi_p(x) = e^{\frac{ipx}{\hbar}}. \qquad (17.17)$$

We can expand its momentum eigenfunction on the basis of position eigenfunction (17.15)

$$\Phi_p(x) = e^{\frac{ipx}{\hbar}} = \int_{-\infty}^{+\infty} dx' e^{\frac{ipx'}{\hbar}} \delta(x - x'). \qquad (17.18)$$

And we would need an infinite number of delta functions to express the wave form, since the Dirac delta is a localized function, while a wave is spread out throughout space. In harmony with the Heisenberg principle (17.13), the definite momentum ($\Delta p = 0$) entails indefiniteness in position ($\Delta x = \infty$), and definite position ($\Delta x = 0$) implies indefiniteness in momentum ($\Delta p = \infty$).

In summary, definitions (17.8), (17.9), (17.10) and (17.11), property (17.12), the uncertainty principle (17.13), de Broglie equation (17.14), and the wavefunctions Ψ and Φ are logically compatible and complete one another.

17.3.7 *Sets of consistent notions* – The statuses $\xi^{(\mathrm{D})}$ and $\xi^{(\mathrm{I})}$ own opposite features. On the one hand, there are integer values of probability, ξ is embodied by a condensed portion of energy/matter and is local. On the other hand, there are decimal values of probability, ξ spreads in the space and is not local (Figure 17.3). The present theory identifies and unifies the contrary aspects of particles and waves by means of the concept of probability.

17.3.8 *Completing the descriptions of quanta* – The theorems of large numbers and a single number prove that probability has

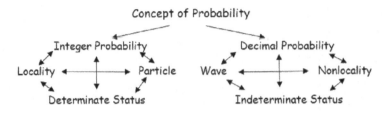

Figure 17.3. Conceptual map of quantum dualism.

different properties depending on the quantity of the elements under scrutiny; thus, we have two different waves to consider.

Definition 17.3. *Assuming* $(n \to \infty)$ *and* $(n = 1)$, *we have respectively*

$$\xi_\infty^{(I)} = A \ stream \ of \ wavelets. \tag{17.19}$$

$$\xi_1^{(I)} = A \ wavelet \ or \ single \ wave. \tag{17.20}$$

In principle, the single particle can occupy a precise position, whereas statistical methods are needed to calculate a group of particles, hence they fall into two classes.

Definition 17.4. *Assuming* $(n \to \infty)$ *and* $(n = 1)$, *we obtain in the order*

$$\xi_\infty^{(D)} = A \ swarm \ of \ particles. \tag{17.21}$$

$$\xi_1^{(D)} = A \ particle. \tag{17.22}$$

It is not difficult to recognize that $\xi_\infty^{(D)}$ describes, for example, a gas of electrons and $\xi_1^{(D)}$ formalizes a proton flying through the space or trapped in a position.

The nature of waves turns out to be more challenging. Definitions (17.19) and (17.20) say that the radiation $\xi_\infty^{(I)}$ is made by countless wavelets, and from (12.24), we obtain

$$\xi_\infty^{(I)} = ((\xi_{11}^{(I)}, \xi_{12}^{(I)}, \xi_{13}^{(I)}, \ldots)). \tag{17.23}$$

Both (17.19) and (17.20) are qualified by *generic decimal values of P* which are objective and pertain to the physical domain (Sections 12.7.6), thus one easily concludes:

Property 17.2

Quantum waves are real.

The left and right sides of (17.23) are equal in nature but greatly differ from the testability viewpoint. The theorems will demonstrate the heavy limitations of experiments in QM.

17.4 Quanta in Movement

The indeterminate statuses of ξ must be studied in the physical environment where they can behave naturally and are not influenced by external factors that distort their conduct. In point of logic, this situation is symmetrical to the movement of a body unaffected by external forces and governed by the first law of CM.

17.4.1 In detail, we assume:

(**X**) *Quanta fly in the Euclidean space Σ and are not subject to any loss of energy; nor are they altered by effects such as spinning, relativity, entanglement and others.*

The phenomenon conforming to requirement (**X**) is known as *free motion*. It consists of the movement of one or more quanta from one place to another place. *The free flight lasts as long as assumption* (**X**) *remains true*; when $\xi^{(I)}$ cedes energy or any physical interaction occurs, the motion is no longer free or even ceases completely. Factually, the measurement process is the most common factor interrupting assumption (**X**); other physical mechanisms can affect $\xi^{(I)}$ and result in the free motion winding up (see Unit III).

17.4.2 This chapter deals with the free motion $\mathbf{E}^{(\xi)}$ that does not obey deterministic mechanical laws. The source i causes ξ to change position in an unpredictable manner (Zeilinger, 2005)

$$\mathbf{E}^{(\xi)} = (i, r; e) = (Emission, Movement; Output \ \xi). \qquad (17.24)$$

The theorems will provide details about $\mathbf{E}^{(\xi)}$.

In conformity with the concepts introduced in Section 16.1, the movement lasts the time interval T1 and finishes at t_ξ when ξ relinquishes its energy. The interval T2 lasts only one instant because of the destructive measurement process which usually causes the end of the flight

$$T1 = (0, t_\xi],$$

$$T2 = t_\xi. \tag{17.25}$$

17.4.3 These introductory tenets provide the first theoretical support for the 'big measurement problem', which could be expressed as follows:

What makes a measurement a measurement?

Speaking in general, the measurement process collects information from the measured object. This operation can be accomplished provided that the input of the measurement device conforms to the *distinguishability property,* which is the defining property of information (Rocchi and Gianfagna, 2005; Rocchi, 2010, 2012, 2016). Only distinct elements enable the accomplishment of measuring tasks, and make measurement a measurement.

17.5 Theoretical Predictions

Let us examine the conduct of ξ during the time intervals T1 and T2.

17.5.1 *Initial conditions* – TIC holds that if $\mathbf{E}^{(\xi)}$ is random, then the outcome is also random in T1, formally we have

$$\xi_\infty = \xi_\infty^{(I)}, \quad 0 \leq t < t_\xi. \tag{17.26}$$

$$\xi_1 = \xi_1^{(I)}, \quad 0 \leq t < t_\xi. \tag{17.27}$$

TIC says that the quanta involved by the aleatory event $\mathbf{E}^{(\xi)}$ are waves no matter whether simple or multiple, massive or massless. This conclusion is consistent with de Broglie's thought that hypothesized that electrons and other discrete bits of matter, which until then had been conceived only as material corpuscles, also have wave properties.

17.5.2 *Continuous and discontinuous behaviors* – If (17.26) is true, the theorem of continuity implies that ξ_∞ remains in the indeterminate status during T2

$$\xi_\infty = \xi_\infty^{(I)}, \quad 0 \leq t. \tag{17.28}$$

Under the hypothesis (17.27), the theorem of discontinuity proves that the indeterminate outcome changes in T2 and becomes certain

$$\xi_1^{(I)} \rightarrow \xi_1^{(D)}, \quad t = t_\xi. \tag{17.29}$$

In essence, the radiation $\xi_\infty^{(I)}$ can be measured; instead $\xi_1^{(I)}$ collapses and transmutes when the free motion ends, that is, when $\xi_1^{(I)}$ absorbs or cedes energy. This ordinarily occurs when $\xi_1^{(I)}$ is measured; therefore, (17.29) proves that the wavelet as such cannot be measured.

TC and TD regard also the two sides of (17.23), which will be discussed in the following pages.

17.5.3 *Superposed statuses* – Let us look more carefully at the behavior of the single wave when a physicist seeks to measure it using N sensors.

TD demonstrates that a sensor interfering with the wavelet perceives an impulse of energy and causes the free motion to terminate. More precisely, the sensor #1 placed at $p_1 = (x_1, y_1, z_1)$ of Σ is able to detect the first potential outcome $\xi_{11}^{(I)}$ of $\mathbf{E}_1^{(\xi)}$. If the operator places the sensor #2 at $p_2 = (x_2, y_2, z_2)$, he determines the second potential outcome $\xi_{12}^{(I)}$ and so forth. When the operator arranges N sensors, he determines N possible alternative outcomes $\xi_{11}^{(I)}, \xi_{12}^{(I)}, \ldots, \xi_{1N}^{(I)}$ which make the *sample space* Ω of the free motion. All of them are virtually available in T1 and overlap in compliance with definition (16.17)

$$\langle \xi_{11}^{(I)}, \xi_{12}^{(I)}, \ldots, \xi_{1N}^{(I)} \rangle, \quad t < t_\xi. \tag{17.30}$$

Basically, N potential outcomes of the quantum experiment rigidly depend on the available sensors.

17.5.4 *Reduction of statuses* – The theorem of reduction predicts that if the single event has N virtual results and winds up, the superposing potential outcomes (17.30) drop-down to just one deterministic outcome

$$\langle \xi_{11}^{(I)}, \xi_{12}^{(I)}, \xi_{13}^{(I)} \cdots \xi_{1j}^{(I)} \cdots \xi_{1N}^{(I)} \rangle \to \xi_{1j}^{(D)},$$

$$j = \text{any of } 1, 2 \ldots, N; \ t = t_\xi. \tag{17.31}$$

This means that $\xi_{11}^{(I)}, \xi_{12}^{(I)} \ldots$ are the virtual outcomes of which only one survives and becomes a particle, while the remaining potential results die. Using (16.19a) and (16.19b), we spell out this special mechanism

$$[0 < P[\xi_{1j} \in (x, y, z)] < 1] \to [F[\xi_{1j} \in (x, y, z)] = 1],$$

$$j = \text{any of } 1, 2, \ldots, N; \quad t = t_\xi. \tag{17.32a}$$

$$[0 < P[\xi_{1k} \in (x, y, z)] < 1] \to [F[\xi_{1k} \in (x, y, z)] = 0],$$

$$k = 1, 2 \ldots (j - 1), (j + 1), \ldots, N; \ t = t_\xi. \tag{17.32b}$$

The change of $\xi_{1j}^{(I)}$, which becomes a particle, and the dissolution of the other potential outcomes are two mechanisms consistent with the quantization principle [1] which states that $\xi_1^{(I)}$ constitutes an indivisible portion of energy and thus it cannot be partitioned.

In summary, there are three theorems which complement one another. TSN proves how the exact probability amplitude of the wavelet cannot be tested, TD explains that $\xi_1^{(I)}$ collapses if an operator tries to measure it, and TR demonstrates the manner in which a wavelet becomes a particle out of N potential outcomes.

17.5.5 *Irreversibility* – The theorem of discontinuity demonstrates that the quantum wave becomes a particle at the end of the flight. The theorem of irreversibility specifies that the transform (17.29) cannot revert

$$P[[\xi_1^{(D)}] \to [\xi_1^{(I)}]] = 0 \quad t = t_\xi. \tag{17.33}$$

The wave becomes a particle, but the particle cannot become a wave. Various authors recognize (17.33), for example, that the collapse of the wave function in the Copenhagen interpretation is manifestly asymmetric in time. Von Neumann (1955) establishes a mathematical framework in which the notion of wavefunction collapse is described as an irreversible process that von Neumann labels 'Process 1'.

Various physical phenomena do not revert and contribute to define the *irreversibility of time*, such as energy radiation, thermodynamic transition, and causal relationship. They are called '*arrows of time*', and result (17.33), named '*quantum arrow of time*' in the literature, belongs to this group of phenomena (Zeh, 2013).

17.6 Comments on the Predictions

The transformation of the single aleatory outcome, which becomes determined, is the universal phenomenon that occurs in classical and quantum physics. It happens in material and virtual terms (Section 16.3), it involves large and small bodies, and is governed by TIC, TSN, TD, and TR. These theorems help us discuss some questions raised in the literature.

17.6.1 Where was the particle in advance of the collapse?

The first postulate of QM associates a wavefunction with any particle moving in a conservative field of force.

The definitions and theorems of this chapter illustrate a completely different landscape. TIC proves that ξ_1 is a wavelet during T1 and there is no wandering corpuscle in space before the collapse. No particle is going in each direction simultaneously. The wavelet is a discrete amount of energy statistically spreading over the space and not "a particle whose position is uncertain". The above written query simply sounds meaningless here.

17.6.2 What is quantum wave collapse?

The wavelet $\xi_1^{(I)}$ is a portion of energy/matter floating in space which transforms into $\xi_1^{(D)}$, thus the *collapse consists of the condensation of the energy/matter in a point.*

This conversion is also consistent with the quantization principle which holds that ξ constitutes an indivisible portion of energy (and occasionally matter). The single wave cannot dispense its energy at various points of the volume V_W, nor can it give up energy at two or three moments. The quantization principle implies that $\xi_1^{(I)}$ necessarily transforms all at once at a point. The current probability theory and the quantum physics provide coherent answers.

17.6.3 Why does the wavelet implode?

When the wavelet cedes energy, its free motion comes to an end, no matter $\xi_1^{(I)}$ is wholly absorbed or continue to move with different energy. Destructive measurement turns out to be the most common reason for the end of free movement even though it is not unique (Unit III). TD proves that in any case the random outcome becomes determined.

17.6.4 When does the collapse occur?

TD specifies that the random outcome becomes certain at the end of the aleatory event; in particular, the wavelet becomes a particle as soon as its movement is no longer free. This happens when any material factor absorbs the energy of the wavelet or anyway assumption **(X)** is no longer true (Section 17.4.1).

17.6.5 If QM predicts various possible outcomes with given probabilities, why does only one of them appear to us?

The emitted wavelet $\xi_1^{(I)}$ is a non-fractional portion of energy/matter diffused in space, and consequently if $\xi_1^{(I)}$ *collapses at a point, it collapses everywhere* due to the quantization principle.

The soap bubble offers the best visual similitude of this quantum mechanism. In fact, the soap bubble, as soon as it interacts with something, condensates and becomes a drop of water. The bubble offers an illustration of the implosion, but there is a difference. The bubble collapses due to a precise interaction that determines the point of collapse. Instead, $\xi_1^{(I)}$ is a statistical distribution and the location of the collapse cannot be predetermined.

17.6.6 Should we consider the superposed states mentioned above as eigenstates?

The Copenhagen interpretation assumes that the quantum wave $|\Psi\rangle$ – initially in a superposition of the eigenstates $|\varphi_1\rangle, |\varphi_2\rangle, \ldots |\varphi_r\rangle$ – reduces to one eigenstate when it collapses. Despite a certain superficial symmetry, the eigenstates have nothing to do with the potential outcomes $\xi_{11}^{(I)}, \xi_{12}^{(I)}, \ldots, \xi_{1N}^{(I)}$ considered in the present framework.

17.6.7 *Definitional and computational formulas* – Quantum waves are real (Property 17.2), but the present purely probabilistic construction does not furnish the exact shapes of waves. According to Born's rule, the probability of finding a quantum at a given point is proportional to the square of the magnitude of the wavefunction at that point

$$P_{\xi I} = P[\xi \in (x, y, z)] = |\Psi(x, y, z)|^2.$$

The wavefunction is the solution to the Schrödinger equation which provides the probability amplitudes depending on specific physical constraints. However, the theorems of single number, discontinuity, and reduction prove that the precise probability amplitude cannot be controlled in the single case. Wavelets cannot be directly measured and give the impression of being 'ghost waves'.

The definitions and the theorems of this chapter illustrate the meaning of the quantum waves and their nature, when waves can be tested and so forth. On the other hand, the wavefunction provides the probability distributions and quantifies other physical features. The two groups of formulas have far different conceptual origins and demonstrate they to complement each other since the first offers *the services typical of DFs* and the second *serves as CFs*.

17.6.8 All conclusions drawn in the preceding pages must be in accordance with facts, and various experiments – already published in the literature – will be used to verify the presence of waves and particles. In practice, we shall confine attention to fluctuating distributions of energy/matter and point-like distributions. Other physical parameters will be overlooked as explained above.

UNIT III – EXPERIMENTS

17.7 Detecting Quanta

Definitions (17.6) and (17.7) posit frequency as the counterpart of probability; the first, available in the physical reality, corroborates the second, derived from calculations.

17.7.1 The nature of quanta as energy/matter and the spatial probability distributions lead us to understand F as the *empirical intensity of energy/matter per unit volume*, where E_T is the grand total *of emitted energy*

$$I(x.y, z) = \frac{E(x.y.z)}{E_T}.\qquad(17.34)$$

The gauged intensity tells us how much energy is located at each point in space and the validity criterion (12.2) becomes

$$P[\xi \in (x, y, z)] \leftrightarrow I(x, y, z).\qquad(17.35)$$

This means that ξ is most likely in those places where the undulations of the wave are greatest or most intense. The denser the energy/matter, the higher the P and vice versa.

The waved continuous distribution of $E(x, y, z)$ brings evidence of waves. Point-like energy manifests a particle, and many particles corroborate a discrete distribution. In summary, I serves to substantiate continuous and discrete states of quanta as we shall see in next pages.

17.7.2 Physicists employ a large variety of sensors to detect quanta: *photographic plates, fluorescent screens, CCD cameras*, etc. However, the exact probability distribution of the wavelet cannot be *directly tested* since $\xi_1^{(I)}$ transmutes when an operator attempts to control it.

Can physicists overcome this experimental limitation?

The stratagem of subjectivists, who assign a personal meaning to P, is not viable for two reasons.

First, we are investigating material phenomena, and we must provide ontological answers to ontological problems. If we supply an epistemic answer to a physical question, we might state something nonsensical and make a category mistake (Section 8.3.3).

Second, scientists employ two experimental techniques to verify theoretical predictions. *Direct testing* gauges the parameter X that the research-team means to investigate; *indirect testing* does not check X but a secondary property connected to X in a way. The restrictions demonstrated by the aforementioned theorems do not exclude indirect methods of testing, and the second technique can demonstrate that the wavelet is not a 'ghost wave' but is real. All this matches with countless works of experimentalists who empirically came up with special stratagems. For example, physicists employ slits, biprisms, semi-transparent mirrors and other devices to recognize the wave state of the incoming single quantum. Moreover, these devices do not absorb energy and merely modify the probability distribution of the incoming $\xi_1^{(I)}$. The secondary waves, created by these instruments, maintain the indivisible unity in consequence of the quantization principle and allow to create various effects for testing purposes.

17.8 Wave Emission

The free movement begins with the emission of quanta, which does not comply with deterministic laws

$$\mathbf{E}^{(\xi)} = (i, r; e) = (Emission,\ Movement;\ Output\ \xi). \qquad (17.24)$$

17.8.1 *Experiments with strong and weak beams* – Let us canvass the characteristics of emission:

- A wide assortment of devices complies with the features of the component i of (17.24) and satisfies the hypotheses of randomness. For example, the following appliances work as ergodic sources: *thermal emitters* (e.g., *a thermionic tungsten filament, a furnace, etc.*), *radioactive sources, plasma emitters, fluorescent tubes*, and *laser systems*.
- These devices substantiate the separation of (17.19) from (17.20), because usually random emitters cast the *intense (or strong) beam* $\xi_\infty^{(I)}$ or alternatively the *weak beam that conveys a wavelet* $\xi_1^{(I)}$ at a time.
- Definitions (17.19), (17.20) and (17.23) hold that $\xi_\infty^{(I)}$ and $\xi_1^{(I)}$ differ only for the number n, namely, they have different testing aptitudes, while the physical nature does not change. The number n influences the testability of probability and not the essence of the waves. Numerous experiments that reduce (or increase) the time-frequency of emitted quanta present evidence of the physical continuity between $\xi_\infty^{(I)}$ and $\xi_1^{(I)}$. For example, the emitter casts an intense beam, while an *attenuator,* an *absorbing medium,* a *filter,* or another instrument brings down the flow of the emitted waves.

This experimental device passes from the left side to the right side of (17.23), and shows there are no physical jumps. The tuning of emitters is a well-known method adopted in (Boyd *et al.*, 2008; Frabboni *et al.*, 2015; Bach *et al.*, 2013) and many other works.

17.8.2 Quantum scientists usually accept the wave properties of photons because photons are regarded as parts of the electromagnetic wave. Instead, the wave properties of electrons, neutrons, and molecules seem puzzling. This section examines three seminal works with electrons, which open the gallery of experiments adopting indirect methods of testing. Section 17.11.1 will look into the duality property of the largest quanta.

Thomson experiment – In 1927, George Paget Thomson used a cathode tube powered by an induction coil. The source **A** launched a beam of electrons against the thin lamina **B**. The photographic plate **C** exhibited the Airy pattern demonstrating the wave nature of electrons (Navarro, 2010). Thomson employed different materials to make the target, such as celluloid, gold, and aluminum, with which he obtained better and more clear images of electron diffraction (Figure 17.4).

17.8.3 ***Davisson–Germer experiment*** – In the same year, Clinton Davisson and Lester Germer sent an electron beam, thermally emitted from the tungsten ribbon **A**, toward a nickel crystal (Weinert, 2009). They observed how the intensity of reflected electrons varies with the angle θ between the detector and the nickel surface (Figure 17.5). The angular dependence of the reflected intensity

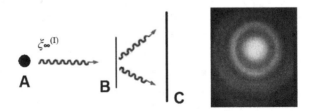

Figure 17.4. Schematic representation of the experiment of G.P. Thomson (left); A circular diffraction pattern [From (Thomson, 1928)] (right).

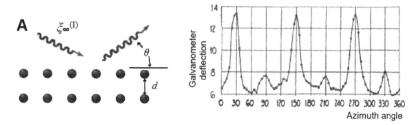

Figure 17.5. Diffraction peaks from the Davisson–Germer experiment (1939) (right-hand side), schematic of the experiment set-up (left-hand side).

is in accordance with *Bragg's law*

$$n\,\lambda = 2\,d\sin\theta \tag{17.36}$$

where n is an integer and d is the distance between the atomic layers. The Bragg law demonstrates that flying electrons are waves – in conformity with the present theory – and their phase shift causes constructive and destructive interferences. Davisson and Germer obtained the following wavelength from (17.36):

$$\lambda = 1.66 \cdot 10^{-10}\,\text{m}. \tag{17.37}$$

The electrons, cast over the target, were accelerated by the potential of $54\,\text{eV}$, and the operators got the wavelength predicted by de Broglie using this value of energy:

$$\lambda_B = \frac{h}{p} = \frac{h}{\sqrt{2mE}} = 1.67 \cdot 10^{-10}\,\text{m}. \tag{17.38}$$

The reader can see how (17.37) and (17.38) turn out to be in good accordance.

17.8.4 *Kapitsa–Dirac experiment* – In 1933, Pyotr Kapitsa and Paul Dirac predicted that electrons should be diffracted by a standing light wave, however, the experiment became feasible much later, only after the invention of the laser (Freimund *et al.*, 2001).

The experimental set is equipped with the stationary laser wave **B,** while the weak beam of electrons crosses it (Figure 17.6 left). The ray **B** must be 'thin' enough so that the interaction time of the electron with the light field has to be sufficiently short in duration and the motion of the electron can be neglected with respect to

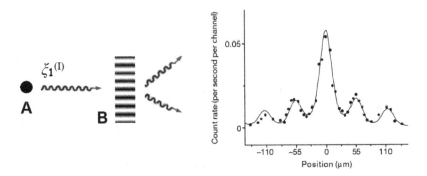

Figure 17.6. Schematic of the Kapitza–Dirac experiment (left-hand side); Diffraction pattern (right-hand side) [From (Freimund *et al.*, 2001)].

the light field (*Raman–Nath regime*). The fringes detected after the electrons have passed through **B** demonstrate that they are waves (Figure 17.6 right). The effect is analogous to the diffraction of light by a grating, but with the roles of the wave and matter reversed.

The experiments just mentioned played a key role in the history of QM and here they substantiate the theorem of initial conditions.

17.9 The Double Slit Experiment

The double slit experiment is often deemed the real, pedagogically clean, and fundamental quantum experiment (Figure 17.9 left-hand side). It has a noteworthy position in the literature and corroborates various theorems presented in this book.

Let us look into the versions of the experiment which use an intense beam and a dim beam, respectively.

▶ When **A** emits a strong beam of photons (Scheider, 1986), or electrons (Lichte, 1988), or other kinds of quanta, then the detector-screen **S** exhibits a continuous pattern (Figure 17.7).

▶ When **A** emits a single quantum, correspondingly the screen **S** shows one dot (Frabboni *et al.*, 2015; Bach *et al.*, 2013; Harada *et al.*, 2018). The greater the number of quanta sent one by one, the more clearly they create a discrete pattern on the screen (Figures 17.8 and 17.11).

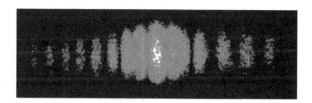

Figure 17.7. Continuous interference and diffraction pattern created by the intense laser light through two slits [From (Changsug *et al.*, 2010)].

Figure 17.8. Progressive discrete pattern created by a weak beam of electrons through two slits [From (Harada *et al.*, 2018)].

Various interpretations of these results have been proposed, yet none seems satisfactory, and Feynman labeled the two slits experiment as "the central mystery of quantum mechanics". The current structural theory yields an entirely novel interpretation.

17.9.1 *Interpretation of the experiment with the intense beam* – Let the incoming wave is plane for the sake of simplicity; the continuous *interference and diffraction pattern* (IDP) created by two slits is described by the following classical formula (Knobel, 2000):

$$I_\xi(\theta) = I_{\xi IN} \cos^2 \alpha_\theta \left(\frac{\sin \beta_\theta}{\beta_\theta} \right)^2 \qquad (17.39)$$

where $I_{\xi IN}$ is the on-axis intensity of the incident plane wave; α_θ and β_θ are defined as follows:

$$\alpha_\theta = (\pi a/\lambda) \sin \theta; \quad \beta_\theta = (\pi b/\lambda) \sin \theta. \qquad (17.40)$$

The symbol λ denotes the wavelength, the constant a is the distance between the slits, and b is the width of each slit (Figure 17.9). The trigonometric functions (17.40) qualify two overposing effects (Figure 17.10):

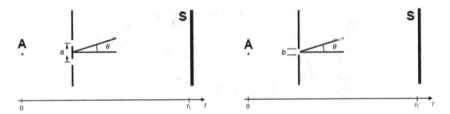

Figure 17.9. Schematic representations of the double slit experiment (left-hand side) and the single-slit experiment (right-hand side).

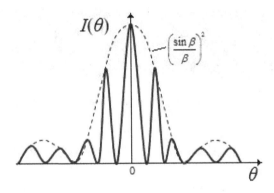

Figure 17.10. Pattern with overposing interference and diffraction effects.

- $(\cos^2 \alpha_\theta)$ = The *interference* resulted from the two waves exiting the slits that overlap.
- $(\sin \beta_\theta / \beta_\theta)^2$ = The *diffraction* caused by the narrow width of the slits.

The continuous spectrum brings evidence that the wave stream $\xi_\infty^{(I)}$ moves in the segment $(\mathbf{A}, \mathbf{S}]$ during the time interval T1 and agrees with TIC

$$\xi_\infty = \xi_\infty^{(I)}, \quad 0 \le t < t_\xi. \tag{17.41}$$

The IDP recorded by $\mathbf{S}$ also indicates that the wave stream remains at time t_ξ – the destructive measurement process causes T2 to last only one instant – and we get

$$\xi_\infty = \xi_\infty^{(I)}, \quad t = t_\xi. \tag{17.42}$$

Equation (17.41), together with (17.42), proves that the intense beam keeps the indeterminate status during the intervals T1 and T2 and corroborates the theorem of continuity.

17.9.2 *Interpretation of the experiment with the weak beam* –
When **A** emits a single quantum, the screen exhibits one spot which brings evidence of one particle in T2

$$\xi_1 = \xi_1^{(D)}, \quad t = t_\xi. \tag{17.43}$$

When the operator repeats the experiment several times, two effects occur simultaneously: the first is about the single wave and the second is about the stream of wavelets.

I] The quanta emitted one by one cannot interact with one another because of the wide time-space separation interposed between them, and physicists conclude that each incoming quantum interferes with itself. That is to say, there is a wavelet in T1

$$\xi_1 = \xi_1^{(I)}, \quad 0 \le t < t_\xi. \tag{17.44}$$

This conclusion conforms with TIC proving that random events bring forth random outcomes. In the present case, the outcomes are wavelets flying in the time interval T1.

Joining (17.43) with (17.44), we obtain the collapse of the single wave that corroborates the theorem of discontinuity

$$\xi_1^{(I)} \to \xi_1^{(D)}, \quad t = t_\xi. \tag{17.29}$$

In conclusion, effect **I** is due to TIC, TD, and principle [1].

II] Definition (17.23) says that in the long run several incoming wavelets make the radiation $\xi_K^{(I)}$

$$\xi_K^{(I)} = ((\xi_{11}^{(I)}, \xi_{12}^{(I)}, \xi_{13}^{(I)}, \ldots, \xi_{1K}^{(I)})), \quad K \to \infty. \tag{17.45}$$

Empirical data shows how the growing number of emitted wavelets betters the contrast of the discrete pattern in **S**. The larger the K, the more clearly the discrete spectrum comes to sight, and substantiates TLN proving that the greater the number of trials, the closer the empirical intensity approaches the calculations. TLN directly implies that the discrete IDP is caused by $\xi_K^{(I)}$.

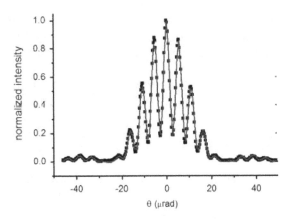

Figure 17.11. Interference and diffraction pattern. created by a dim beam of electrons through two slits [From (Frabboni *et al.*, 2015)].

Finally, the pattern on **S** brings evidence that the stream $\xi_K^{(I)}$ remains in the wave status in T1 and T2, and this detail corroborates the theorem of continuity.

In summary, effect **I** regards the single wavelet that interferes with itself and collapses due to the measurement process. Effect **II** regards the overall emission, which remains indeterminate in T1 and T2, and creates the IDP.

The theorem of initial conditions proves that the incoming quanta are wavelets. The theorem of discontinuity predicts the collapse of $\xi_{1j}^{(I)}$ ($j =$ any of $1, 2 \ldots$) in **I**. The theorems of large numbers and continuity ensure that $\xi_K^{(I)}$ can be verified in **II**. The effects **I** and **II** occur in parallel and are consistent with the corollary of separation (Section 13.2). Basically, the operator controls the left and the right sides of (17.45) in different manners.

No doubt that the double slit experiment with the weak beam turns out to be rather complex due to the simultaneous effects **I** and **II**, each one involving various theorems.

17.10 The Single Slit Experiment

The beam of quanta is passed through one slit of finite width and the detector-screen **S** exhibits the results (Figure 17.9). Numerous tests have been performed with photons (Zhou *et al.*, 2018), electrons (Bach *et al.*, 2013), and neutrons (Zeilinger *et al.*, 1988) which provide indisputable evidence of the central peak and the secondary peaks of the pattern on **S**. The diffraction patterns are continuous or discrete depending on the incoming beam that can be strong or weak in the order (the analysis of the outcomes, symmetrical to that of Sections 17.9.1 and 17.9.2, is omitted).

It is important to underscore how the spectrum generated from the single slit has a different shape than that created by two slits and creates more difficulties for scientists who have sometimes drawn misleading conclusions.

The spectrum generated by one slit depends only on the *diffraction* effect, and the following equation, lacking the interference term, regulates the diffraction pattern (DP):

$$I_\xi(\theta) = I_{\xi IN} \left(\frac{\sin \beta_\theta}{\beta_\theta} \right)^2. \tag{17.46}$$

When θ approaches zero, the following limit is one:

$$\lim_{\beta \to 0} \left(\frac{\sin \beta}{\beta} \right) = 1. \tag{17.47}$$

Putting this limit value in (17.46), we can conclude that the intensity of the central peak I_0 approximates the incident wave

$$I_0 \approx I_{\xi IN}. \tag{17.48}$$

Most of the intensity is concentrated in the central maximum, while the bands on either side have negligible intensities. More precisely the first peak placed aside I_0 satisfies the following equation (demonstration is omitted):

$$I_1 = 0.04 \, I_0. \tag{17.49}$$

The second maximum, aside the central peak, is two and a half times lower than I_1

$$I_2 = 0.016 \, I_0. \tag{17.50}$$

Figure 17.12. Diffraction pattern created by a weak beam of electrons through one slit [From (Bach *et al.*, 2013)].

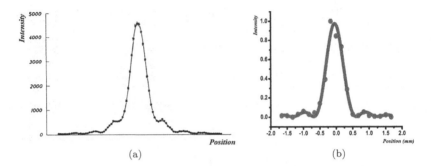

Figure 17.13. Diffraction patterns created through one slit. **(a)** by a weak beam of neutrons [From (Zeilinger *et al.*, 1988)]; **(b)** by a weak beam of photons [Fom (Zhou *et al.*, 2018)].

The patterns reproduced in Figures 17.12 and 17.13 make evident how the central peak has far greater intensity than those on either side and corroborate the present interpretation.

The difficult visualization of the lateral fringes had great influence in the history of QM since Feynman (1965) noted how two slits create a typical wave IDP, but if one slit is closed, the operator obtains a distribution of dots without fringes. Feynman illustrates his thought experiment using Figure 17.14, where P_1 and P_2 are the distributions created by the two beams of electrons that pass through each single slit. The profiles of P_1 and P_2 call to mind the Gaussian distribution and seem to prove that electrons behave as particles when they pass through a single slit. Bach and his colleagues (Bach *et al.*, 2013) have factually conducted the experiment envisioned by Feynman. Bach's team physically moved a mask across the slits so that each could be individually closed, or both could be open. The upshots brought evidence of the feeble lateral fringes that the single slit originates (Figure 17.12).

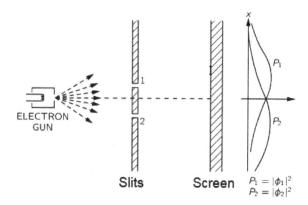

Figure 17.14. Feynman's illustration of the mind-experiment with one slit open at a time [From (Feynman, 1965)].

In conclusion, the single slit experiment brings evidence that the incoming quanta make a diffraction pattern and therefore they are waves. The idea that **A** emits a wave in front of two slits and a particle in front of one slit turns out to be impossible to understand because it does not correspond to the facts.

17.11 Variants of the Double Slit Experiment

Researchers have devised several versions of the double slit experiment to pursue different scopes.

17.11.1 *Experiments with large molecules* – The double slit experiment with large molecules yields four principal conclusions.

(A) The wavelength of de Broglie (1950) is inversely proportional to the mass of ξ

$$\lambda_B = \frac{h}{p} = \frac{h}{mv}. \qquad (17.14)$$

Therefore, the very short λ associated with a 'large' mass is much harder to observe compared to the wavelength of the smaller corpuscles.

Significant improvements in interferometric resolution techniques have enabled scientists to observe the patterns created by *neutrons* (Zeilinger *et al.*, 1988), *helium atoms* (Carnal

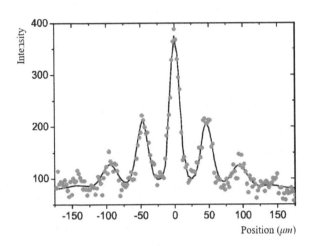

Figure 17.15. Interference and diffraction pattern created by a weak beam of C_{60} with two slits [From (Nairz *et al.*, 2003)].

and Mlynek, 1991), C_{60} *fullerenes* (Arndt *et al.*, 1999; Nairz *et al.*, 2003) (Figure 17.15), *Bose–Einstein condensates* (Andrews *et al.*, 1997), and *biological molecules* (Hackermüller *et al.*, 2003). Eibenberger and colleagues (2013) have created fringes of interference by launching molecules of about 10^4 atomic mass units, roughly about 5000 protons and 5000 neutrons. Physicists have also exploited the Kapitza–Dirac effect to check the wave state of atoms (Schaff *et al.*, 2014). All the experiments present interference patterns which can be interpreted by means of TIC, namely, *atoms and molecules, coming out of the ergodic sources, are waves.*

(B) The quantization principle [1] and the duality principle [2] formulate the essential characteristics of quanta in the present framework. They state that *a body that takes on discrete values of energy/matter, and exhibits particle and wave behaviors, is a quantum.*

It is universally recognized that atoms and molecules are discrete portions of matter and conform to [1]. The abovementioned experiments bring evidence that they are also in line with [2], so it is reasonable to conclude that *atoms and molecules are quanta.*

(C) The greater the m in (17.14), the shorter the λ_B, that is, the wavelength associated to a 'large' body becomes negligible

$$\lim_{m \to \infty} \lambda_B = 0. \qquad (17.51)$$

This limit has noteworthy value. First, it disproves the fable that everything, even a large body, could have an associated wavefunction. It denies the idea that QM is 'universal' and can be used to describe systems at all scales. Second, it is reasonable to forecast that the more the testing techniques will progress, the better the limit (17.51) will be controlled. Experimentalists will throw corpuscles of increasing size against two slits so that any interference pattern will fade away and the quantum dualism will no longer hold. In this manner, the present theory, together with the scheme of de Broglie, indicates how to find out the boundary between QM and CM.

(D) The theorems proved in this book cross classical and quantum physics, and, in this manner, they also provide an answer to the correspondence problem raised by Bohr (Section 17.1.4). Probability establishes continuity and symmetry between QM and CM, while criterion (17.51) formulates the boundary between the two physical realms.

17.11.2 'Which way' experiments – Several scientists assume that the photons, electrons, etc. emitted by the ergodic source **A** are particles, so they try to discover the slit crossed by the supposed particle. The 'which way' or 'welcher Weg' problem does not make sense from the present perspective since TIC proves that the quantum emitted by **A** is a wavelet; nonetheless, the attempts made so far turn out to be instructive and strengthen the predictions of the present chapter.

Various techniques have been excogitated to detect the slit crossed by the imagined particle (Busch and Jaeger, 2009; Bucks *et al.*, 1998; Dürr *et al.*, 1998; Kincaid *et al.*, 2016; Xiao *et al.*, 2019). Let us overlook the details and pay attention to the simplest experimental setting (Figure 17.16) that is equipped with one sensor or two sensors **Q** placed behind the slits in order to recognize the hypothetical crossed slit.

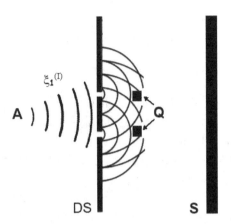

Figure 17.16. Schematic of the 'which way' experiment.

The theorem of initial conditions proves that a wavelet crosses the slits, so they generate two secondary waves. When a secondary wave clashes against **Q**, it cedes energy and the overall wavelet collapses because the secondary waves make an indivisible unit in consequence of the quantization principle [1]. When the operator places two sensors, the secondary wave is registered in one detector, not in both, and this detail confirms the present interpretation. The measuring interaction with **Q** results in the wavelet collapsing and, of course, the screen **S**, which is placed behind **Q**, does not detect any signal. More precisely, the more secondary waves are intercepted by **Q**, the more the pattern on **S** is faint. This theoretical prediction concords with experimentalists who claim: "The more information one gets about the crossed slit, the more the spectrum of interference fades or vanishes."

In summary, 'which way' experiments show that the untroubled wavelet collapses in **S**; instead, when a secondary wave hits the sensor **Q**, the wavelet collapses in **Q**, and **S** does not register any dot. The present theoretical framework demonstrates:

- $\xi_1^{(I)}$ inputs the slits and not $\xi_1^{(D)}$.
- The signal intercepted by **Q** does not provide information about the crossed slit, it notifies the secondary wave which interferes randomly with **Q**.

17.11.3 *Experiment with polarizers* – A special version of the double slit experiment employs the laser **A** emitting single photons (Dimitrova and Weis, 2011). The experiment consists of three stages:

1. Photons incident on two slits create IDP on the detector screen **S**.
2. The polarizers HV with vertical–horizontal orientations are set immediately behind DS (Figure 17.17) and the pattern disappears in **S** (Figure 17.18 top).
3. When the polarizer ER, oriented at ±45°, is added before **S**, the interference reappears (Figure 17.18 bottom).

Various authors believe that the polarizers HV convey information about the path taken by the incoming particles (Mittelstaedt *et al.*, 1987; Hillmer and Kwiat, 2007; Rueckner and Peidle, 2013). The second polarizer ER would 'erase' this information, allowing the diffraction pattern to reappear. Some theorists complete the explanation by appending considerations about the complementarity principle, the uncertainty principle, etc. The present theory follows a far different line of reasoning to interpret this experiment.

TIC proves that each incoming photon is a wavelet and not a particle, hence the polarized images can all be explained on the basis of the wave mechanics. The incoming wavelet crosses the slits and

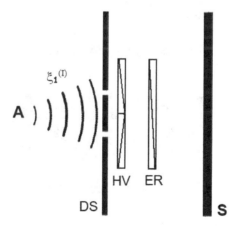

Figure 17.17. Schematic of the double slit experiment using polaroid filters.

creates two coherent secondary waves that interfere and make a discrete pattern on the screen (step **1**). When the operator places the polarizers HV, there can be two possibilities:

A. Either the horizontal polarizer H or the vertical polarizer V covers both slits, and the IDP is preserved on **S**. In fact, the secondary waves undergo an identical polarization process and remain coherent, albeit at reduced intensity compared with no filter.

B. The polarizer H covers one slit and V covers the other (Figures 17.18 top); the diffraction pattern disappears completely as the secondary waves have orthogonal polarized planes (step **2**).

In case **B**, the second polarization ER that is oriented at $\pm 45°$ re-aligns the planes of the secondary waves, which make the fringes on **S** anew (Figure 17.18 bottom) (step **3**). In summary, the three steps of the experiment can be accounted for by TIC and classical optics (Collett, 1993).

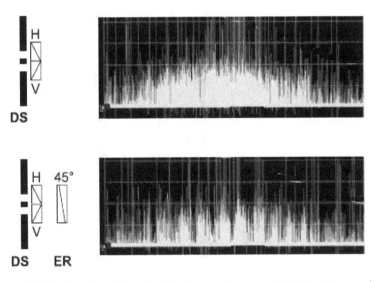

Figure 17.18. Results of the double slit experiment with HV polarizers (top); and with HV+ER polarizers (bottom) [From (Dimitrova and Weis, 2011)].

17.12 Experiments with Biprisms

Scientists have created wave interference patterns using biprisms in the place of slits. In detail:

✓ They have used the optical biprism as a thin double prism placed base to base which has a very small refracting angle. Veit and Solarek (1975) have launched an *intense laser beam* through an optical biprism and created an interference pattern.

✓ They have used the electrical biprism consisting of two parallel grounded plates with the filament F placed between them bisecting the flow of electrons (Figure 17.19). Merli and colleagues (1976), Tonomura and colleagues (1989), and others have driven a *weak beam* of electrons and have observed the discrete pattern **E** on **S**. The progressive construction of the discrete pattern brings further evidence of the simultaneous effects **I** and **II** already discussed in Section 17.9.

17.13 The Einstein Thought Experiment

Einstein presented a thought experiment at the 1927 Solvay Conference, which was later picked up and modified by various authors (Norsen, 2005). Let us examine the simplest version of this experiment.

Suppose the plane wave is incident on a diaphragm with a single aperture, behind which lies a large, hemispherical sensing screen

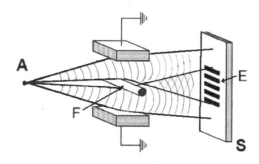

Figure 17.19. Schematic of the experiment with electrical biprism.

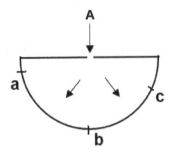

Figure 17.20. Schematic representation of the Einstein thought experiment.

(Figure 17.20). With a sufficiently narrow aperture, the incident quantum wave will diffract, resulting in essentially spherical waves propagating toward the screen. The spherical wave can be detected at points **a**, **b** and **c**. If detector **b** perceives the wave that collapses, how can detectors **a** and **c** be informed about this collapse?

This thought experiment hypothesizes that the wavelet $\xi_1^{(I)}$ can be detected by the receptors **a**, **b**, or **c**, hence $\xi_{1\mathbf{a}}^{(I)}$, $\xi_{1\mathbf{b}}^{(I)}$, and $\xi_{1\mathbf{c}}^{(I)}$ are the potential results that satisfy the superposition hypothesis (17.30)

$$\langle \xi_{1\mathbf{a}}^{(I)}, \xi_{1\mathbf{b}}^{(I)}, \xi_{1\mathbf{c}}^{(I)} \rangle.$$

Suppose **b** perceives the wavelet, the theorem of reduction (17.31) forecasts the following transitions:

$$[0 < P(\xi_{1\mathbf{b}}) < 1] \to [F(\xi_{1\mathbf{b}}) = 1],$$
$$[0 < P(\xi_{1\mathbf{a}}) < 1] \to [F(\xi_{1\mathbf{a}}) = 0],$$
$$[0 < P(\xi_{1\mathbf{c}}) < 1] \to [F(\xi_{1\mathbf{c}}) = 0], \quad t = t. \qquad (17.52)$$

This means that when $\xi_{1\mathbf{b}}^{(I)}$ switches to $\xi_{1\mathbf{b}}^{(D)}$, the statuses $\xi_{1\mathbf{a}}^{(I)}$ and $\xi_{1\mathbf{c}}^{(I)}$ also change; more precisely the wavelet reduces to a particle in **b**, while $\xi_{1\mathbf{a}}^{(I)}$ and $\xi_{1\mathbf{c}}^{(I)}$ vanish and thus the detectors **a** and **c** *do not need to be informed*. The incoming individual $\xi_1^{(I)}$ collapses as a whole in harmony with the quantization principle and the theorem of reduction. This thought experiment reaffirms the image of the wavelet that collapses like a soap bubble.

17.14 Experiments with the Interferometer

The experimental set consists of a Mach–Zehnder interferometer, where the laser **A** emits a weak beam of photons (Grangier *et al.*, 1986; Jacques *et al.*, 2007). The beam-splitter BS_{in} divides the beam into two equal parts. The mirrors M_H and M_K redirect the beams toward the detectors D_H and D_K in the order. Each detector, when it senses a photon, shows a dot and also sends a click. The experiment encompasses two phases.

In the first phase, the optical circuit is open (Figure 17.21 left). Source **A** sends a sequence of single photons; D_H and D_K detect the incoming photons from M_H and M_K in the order, while the phase-shift Φ (indicated with arbitrary origin) is obtained by slightly varying the two arms of the interferometer. The black dots in Figure 17.22 represent the pattern seen by D_H, the gray dots are due to D_K.

In the second phase, the operator closes the optical circuits using the device BS_{out} (Figure 17.21 right) that recombines the beams. The black and gray dots describe the discrete interference patterns created on BS_{out} and viewed by D_H and D_K (Figure 17.23). The clicks of D_H and D_K never occur at the same time.

Quantum scientists still dispute the meanings of the results produced by the two phases of the experiment. John Archibald Wheeler believes that the light somehow "senses" the experimental apparatus through which it will travel and adjusts its behavior to fit by assuming either the wave or the particle state.

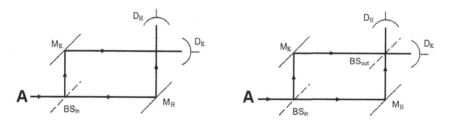

Figure 17.21. The open (left side) and close (right side) configurations of the experiment with Mach–Zehnder interferometer.

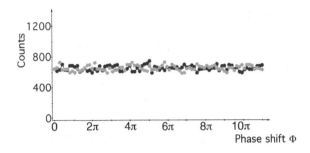

Figure 17.22. Results of the first phase of the experiment [From (Jacques *et al.*, 2007)].

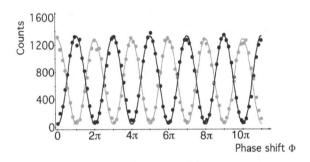

Figure 17.23. Results of the second phase of the experiment [From (Jacques *et al.*, 2007)].

The theorems presented in the previous chapters lead to the following interpretation, which has nothing in common with the literature.

17.14.1 *Interpretation of the first phase* – The first phase presents two effects, which occur simultaneously: the first involves the individual wavelet and the second involves the stream of wavelets.

***A*]** TIC ensures that each individual photon exiting from the laser is a wavelet

$$\xi_1 = \xi_1^{(I)}, \quad 0 \le t < t_\xi. \tag{17.27}$$

The wavelet, after crossing BS_{in}, creates $\xi_{1H}^{(I)}$ and $\xi_{1K}^{(I)}$ that are the potential results of the experiment perceivable by D_H and D_K.

They are superposing virtual outcomes in agreement with definition (17.30)

$$\langle \xi_{1H}^{(I)}, \xi_{1K}^{(I)} \rangle, \quad t < t_\xi. \tag{17.53}$$

The theorem of reduction (17.32a) and (17.32b) proves that only one potential outcome collapses, namely, it is detected, while the other potential outcome vanishes

$$[0 < P(\xi_{1X}) < 1] \rightarrow [F(\xi_{1X}) = 1],$$

$$X = \text{either H or K}; \quad t = t_\xi. \tag{17.54a}$$

$$[0 < P(\xi_{1Y}) < 1] \rightarrow [F(\xi_{1Y}) = 0],$$

$$Y \neq X; \quad t = t_\xi. \tag{17.54b}$$

The clicks which do not overlap corroborate this interpretation.
B] The beam-splitter BS$_{\text{in}}$ creates $\xi_{1H}^{(I)}$ and $\xi_{1K}^{(I)}$ that are coherent secondary waves in addition to being superposed potential outcomes as in (17.53). In the long run, BS$_{\text{in}}$ creates the streams of $\xi_{1H}^{(I)}$ and $\xi_{1K}^{(I)}$, which are statistically balanced, and the registered dots confirm that the streams are symmetrical (Figure 17.22).

17.14.2 *Interpretation of the second phase* – This phase also has two simultaneous effects involving the single wavelet and the stream of wavelets in the order.

1] The theorem of reduction (17.54a) and (17.54b) predicts that a potential outcome – $\xi_{1H}^{(I)}$ or $\xi_{1K}^{(I)}$ – collapses, while the other vanishes. The probes D$_H$ and D$_K$ sense particles, while the clicks never overlap (Figure 17.23), they corroborate this mechanism that is symmetrical to **A]**.
2] This effect prompts two remarks.

- The beam-splitter BS$_{\text{in}}$ creates coherent secondary waves. The device BS$_{\text{out}}$, which works like a semi-transparent mirror, recombines $\xi_{1H}^{(I)}$ and $\xi_{1K}^{(I)}$ that create constructive and destructive interferences. The sensors D$_H$ and D$_K$, which detect two patterns on BS$_{\text{out}}$ because of their distinct perspectives (Figure 17.23), substantiate this interpretation.

    – The large number of incoming photons bring into being $\xi_\infty^{(I)}$; and the discrete patterns shown by BS_{out} belong to wave-stream $\xi_\infty^{(I)}$ in consequence of (17.23). This conclusion is also consistent with TLN and TC that ensure the long-term event can be tested.

In summary, $\xi_{1H}^{(I)}$ and $\xi_{1K}^{(I)}$ behave as superposed potential outcomes in stages A] and 1]; $\xi_{1H}^{(I)}$ and $\xi_{1K}^{(I)}$ behave as coherent secondary waves in stages B] and 2]. This experiment is somewhat complicated due to the simultaneous effects occurring in each phase that are governed by four theorems: TLN, TIC, TC, and TR.

17.15 The Lenard Experiment

Philipp Lenard first investigated the phenomenon of photoelectric emission in a systematic manner (Wheaton, 1978). The core of the experimental apparatus consists of a beam of monochromatic light hitting the cathode of a phototube (Figure 17.24). For a given metal surface, there is a threshold frequency of incident radiation above which photoelectrons are emitted by the cathode. More precisely, the kinetic energy E_K of the emitted electrons depends on the frequency ν of the incoming light

$$E_K = h\nu - W, \qquad (17.55)$$

where W is the minimum energy required to remove an electron from the surface of the material. Einstein's formula, however simple, explains the essence of the photoelectric mechanism, it specifies that each photon cedes energy to an electron that receives sufficient energy to exit from the cathode and move toward the anode. This is

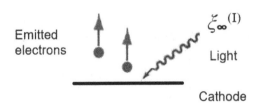

Figure 17.24. Schematic of the Lenard experiment.

consistent with the present theory holding that the intense beam of light (17.23) is made by several individual wavelets

$$\xi_\infty^{(I)} = ((\xi_{11}^{(I)}, \xi_{12}^{(I)}, \xi_{13}^{(I)}, \ldots)). \tag{17.23}$$

The generic wavelet $\xi_{1j}^{(I)}$ ($j = 1, 2, 3, \ldots$) hits randomly an electron; it gives up energy and collapses. The current framework offers two advantages:

- It justifies the compatibility of the incoming electromagnetic radiation $\xi_\infty^{(I)}$ with the photons that operate one by one and factually realize the photoelectric effect.
- It gives an example of how the free flight of the wavelet comes to an end without the intervention of any measurement process.

17.16 Excluded Experiments

The present essay verifies the theoretical predictions under the constraint **(X)** (Section 17.4.1) which requires that all external mechanisms be left out. For this reason, several experiments lie outside the perimeter of the present research such as those which cast spinning particles (Wheaton, 1978), pairs of momentum-entangled photons (Zeilinger, 1999), signal and idler photons (Kim *et al.*, 2000; Scully and Drühl, 1982), Bose–Einstein condensates moved by the gravity field (Manning *et al.*, 2015), experiments with particle accelerators and so forth.

Other tests are alien to **(X)** such as the Hong-Ou-Mandel (1987) experiment that utilizes the probabilistic features of quanta in order to assess the physical properties of fermions and bosons.

UNIT IV – CLOSING

17.17 Conclusion and Perspectives

This chapter, which could be defined as a *probability-based interpretation of quantum mechanics*, begins by discussing the methodological requisites that must be satisfied. It demonstrates that scientists cannot successfully solve the basic quantum issues until modern probability theories are conflicting and cover narrow areas.

The present ample theory of probability does not merely provide mathematical support; it introduces a new vision of the physical world where the tiniest parts of energy and matter either concentrate in spots or spread in space. The particles and waves are described by spatial distributions of probability while the Dirac function illustrates the first, and the squared wavefunction depicts the second.

The present theory underscores the identical nature of $\xi_\infty^{(I)}$ and $\xi_1^{(I)}$ and also the difference between the radiation $\xi_\infty^{(I)}$ which experiments can confirm, and the single wave $\xi_1^{(I)}$ that is untestable. This essay also provides insights about the *soap-bubble effect* typical of the wavelet which, when it collapses in a point, collapses everywhere. This theory can also provide original interpretations of the EPR paradox and the entanglement which will be published later.

Any indeterminate state becomes determinate at the end of the single random event, regardless of whether the event belongs to classical or quantum physics. The theorems create a formal bridge between classical and quantum physics while the theory of de Broglie establishes the boundary between the two physical domains. In summary the structural theory of probability provides precise coherent answers to the following fundamental issues:

✓ Quantum dualism,
✓ Wave collapsing,
✓ Quantum measurement,
✓ Relations between CM and QM.

The explanations also meet intuition and show how the weirdness of QM – emphasized for a long – takes its origin from mathematics and not from physics. At the same time, it makes clear that the following ideas circulating in the literature do not have ground here:

• Wavefunction is definitional,
• Consciousness causes collapse,
• Delayed choice,
• Quantum erasers,

- Retrocausality,
- Hidden variables,
- 'Which way' information,
- Many-worlds interpretation.

This study centers on the free motion of quantum waves, which corresponds to the first law of classical mechanics dealing with a body that is not acted upon by any force. Research should continue to address more complex situations. For example, it would be interesting to investigate the relations with the relativity theory and so forth. It would be highly recommended that some experiments be improved in the future. For example, it would be good to gauge the wavelengths of diffracted beams.

References

Andrews M.R., Townsend C.G., Miesner H., Durfee D.S., Kurn D.M., and Ketterle W. (1997). Observation of interference between two Bose condensates, *Science*, 275(5300), 637–641.

Arndt M., Nairz O., Vos-Andreae J., Keller C., Van der Zouw G., and Zeilinger A. (1999). Wave–particle duality of C60 molecules, *Nature*, 401, 680–682.

Bach R., Pope D., Liou S., and Batelaan H. (2013). Controlled double-slit electron diffraction, *New Journal of Physics*, 15(3), 1–7.

Bell J. (1990). Against 'measurement', *Physics World*, 3(8), 33–40.

Bohr N. (1949). Discussions with Einstein on epistemological problems in atomic physics, In P.A. Schilpp (ed.), *Albert Einstein: Philosopher–Scientist*, pp. 201–241 (Northwestern University Press, Evanston, IL).

Bohr N. (2010). *Atomic Physics and Human Knowledge* (Dover Publications, Chicago).

Boniolo G. (2000). What does it Mean to Observe Physical Reality?, In E. Agazzi and M. Pauri (eds.), *The Reality of the Unobservable: Observability, Unobservability and Their Impact on the Issue of Scientific Realism*, 177–190 (Springer, Berlin; New York).

Boughn S. and Reginatto M. (2013). A pedestrian approach to the measurement problem in quantum mechanics, *European Physical Journal H*, 38, 443–470.

Boyd R.W., Agarwal G.S., Clifford Chan K.W., Jha A.K., and O'Sullivan M.N. (2008). Propagation of quantum states of light through absorbing and amplifying media, *Optics Communications*, 281, 3732–3738.

Bucks E., Schuster R., Heiblum M., Mahalu D., and Umansky V. (1998). Dephasing in electron interference by a "which-path" detector, *Nature*, 391, 871–874.

Busch P. and Jaeger G. (2009). Which-way or Welcher-Weg-experiments, In D. Greenberger, K. Hentschel, and F. Weinert (eds.), *Compendium of Quantum Physics*, pp. 845–851 (Springer, Berlin; New York).

Carnal O. and Mlynek J. (1991). Young's double-slit experiment with atoms: A simple atom interferometer, *Physical Review Letter*, 66, 2689–2692.

Changsug L., Kwangmoon S., Sungmuk L., and Jaebong L. (2010). Fabrication of slits for Young's experiment using graphic arts films, *American Journal of Physics*, 78(1), 71–74.

Colbeck R. and Renner R. (2012). Is a system's wave function in one-to-one correspondence with its elements of reality?, *Physical Review Letter*, 108, 150402.

Collett E. (1993). *Polarized Light: Fundamentals and Applications* (Marcel Dekker, New York).

Cushing J. (1994). *Quantum Mechanics, Historical Contingency, and the Copenhagen Hegemony* (University of Chicago Press, Chicago).

Davisson C.J. and Germer L.H. (1928). Reflection of electrons by a crystal of Nickel, *Proceedings of the National Academy of Sciences of the United States of America*, 14(4), 317–322.

de Broglie L. (1950). *La Mécanique Ondulatoire des Systèmes de Corpuscules* (Gauthier-Villars, Paris).

Dimitrova P.L. and Weis A. (2011). A portable double-slit quantum eraser with individual photons, *European Journal of Physics*, 32(6), 1535–1546.

Dodd J.E. and Gripaios B. (2020). *The Ideas of Particle Physics* (Cambridge University Press, Cambridge).

Dürr S., Nonn T., and Rempe G. (1998). Fringe visibility and which-way information in an atom interferometer, *Physical Review Letter*, 81, 5705–5709.

Eibenberger S., Gerlich S., Arndt M., Mayor M., and Tüxen J. (2013). Matter-wave interference with particles selected from a molecular library with masses exceeding 10000 amu, *Physical Chemistry Chemical Physics*, 15, 14696–14700.

Einstein A. (1909). On the present status of the radiation problem, *Physikalische Zeitschrift*, 10, 185–193.

Einstein A. and Infeld L. (1938). *The Evolution of Physics: The Growth of Ideas from Early Concepts to Relativity and Quanta* (Cambridge University Press, Cambridge).

Feynman R.P. (1965). *The Feynman Lectures on Physics*, Vol. 3, pp. 1–8. (Addison-Wesley, Boston, MA).

Frabboni S., Gazzadi G.C., Grillo V., and Pozzi G. (2015). Elastic and inelastic electrons in the double-slit experiment: A variant of Feynman's which-way set-up, *Ultramicroscopy*, 154, 49–56.

Freimund D., Aflatooni K., and Batelaan H. (2001). Observation of the Kapitza–Dirac effect, *Nature*, 413, 142–143.

Fuchs A. and Peres A. (2000). Quantum theory needs no 'interpretation', *Physics Today*, 53, 70–1.

Ghose P.A. (2002). Continuous transition between quantum and classical mechanics, *Foundations of Physics*, 32, 871–892.

Grangier P., Roger G., and Aspect A. (1986). Experimental evidence for a photon anticorrelation effect on a beam splitter: A new light on single-photon interferences, *Europhysics Letters*, 1(4), 173–179.

Gregg J. (2014). What in the (quantum) world is macroscopic?, *American Journal of Physics*, 82(9), 896–905.

Gudder S.P. (1988). *Quantum Probability* (Academic Press, Cambridge, MA).

Hackermüller L., Uttenthaler S., Hornberger K., Reiger E., Brezger B., Zeilinger A., and Arnd M. (2003). Wave nature of biomolecules and fluorofullerenes, *Physical Review Letter*, 91, 90408.

Harada K., Akashi T., Niitsu K., Shimada K., Ono Y.A., Shindo D., Shinada H., and Mori S. (2018). Interference experiment with asymmetric double slit by using 1.2-MV field emission transmission electron microscope, *Scientific Reports*, 8(1), 1–10.

Heisenberg W. (1927). The physical content of quantum kinematics and mechanics, In J.A. Wheeler and W.H. Zurek (eds.), *Quantum Theory and Measurement*, pp. 62–84 (Princeton University Press, Princeton, NJ).

Hillmer R. and Kwiat P. (2007). A do-it-yourself quantum eraser, *Scientific American*, 296(5), 90–95.

Hong C.K., Ou Z.Y., and Mande, L. (1987). Measurement of subpicosecond time intervals between two photons by interference, *Physics Review Letters*, 59(18), 2044–2046.

Jacques V., Wu E., Grosshans F., Treussart F., Grangier P., Aspect A., and Roch J.F. (2007). Experimental realization of Wheeler's delayed-choice gedanken experiment, *Science*, 315 (5814), 966–968.

Khrennikov A. (2016). *Probability and Randomness: Quantum versus Classical* (Imperial College Press, London).

Kim Y.H., Yu R., Kulik S.P., Shih Y.H., and Scully M.O. (2000). A delayed choice quantum eraser, *Physical Review Letter*, 84, 1–4.

Kincaid J., McLelland K., and Zwolak M. (2016). Measurement-induced decoherence and information in double-slit interference, *American Journal of Physics*, 84, 522, arXiv:1606.09442.

Knobel R. (2000). *An Introduction to the Mathematical Theory of Waves* (American Mathematical Soc. Publ., Providence, RI).

Kümmerer B. and Maassen H. (1998). Elements of quantum probability, In R.L. Hudson and J.M. Lindsay (eds.), *Quantum Probability Communications*, Vol. X, pp. 73–100 (World Scientific, Singapore).

Landsman N.P. (2007). Between classical and quantum, *Handbook of the Philosophy of Science*, Vol. 2: *Philosophy of Physics*, Part A, 417–553.

Lichte H. (1988). Electron interferometry applied to objects of atomic dimensions, In D.M. Greenberger (ed.), *New Techniques and Ideas in Quantum Measurement Theory*, p. 175 (Academy of Science, Washington, DC).

Manning A.G., Khakimov R.I., Dall R.G., and Truscott A.G. (2015). Wheeler's delayed-choice gedanken experiment with a single atom, *Nature Physics*, 11(7), 539–542.

Merli P.G., Missiroli G.F., and Pozzi G. (1976). On the statistical aspect of electron interference phenomena, *American Journal of Physics*, 44, 306–307.

Mittelstaedt P., Prieur A., and Schieder R. (1987). Unsharp particle-wave duality in a photon split-beam experiment, *Foundations of Physics*, 17(9), 891–903.

Myrvold W. (2016). Philosophical issues in quantum theory, *Stanford Encyclopedia of Philosophy*, Available at https://plato.stanford.edu/entries/qt-issues/.

Nachtmann O. (2012). *Elementary Particle Physics: Concepts and Phenomena* (Springer, Berlin; New York).

Nairz O., Arndt M., and Zeilinger A. (2003). Quantum interference experiments with large molecules, *American Journal of Physics*, 71(4), 319–325.

Navarro J. (2010). Electron diffraction chez Thomson: Early responses to quantum physics in Britain, *The British Journal for the History of Science*, 43(2), 245–275.

Norsen T. (2005). Einstein's boxes, *American Journal of Physics*, 73(2), 164–176.

Peierls R. (1991). In defence of 'measurement', *Physics World*, 4, 19–20.

Qureshi T. and Zini R. (2012). Quantum eraser using a modified Stern-Gerlach setup, *Progress of Theoretical Physics*, 127(1), 71–78.

Rae A.I.M. (2004). *Quantum Physics: Illusion or Reality?* (Cambridge University Press, Cambridge).

Rocchi P. (2010). Notes on the essential system to acquire information, *Quantum Information and Entanglement* special issue of *Advances in Mathematical Physics*, 2010(480421).

Rocchi P. (2012). *Logic of Analog and Digital Machines*, 2nd revised edition (Nova Science Publishers, Hauppauge, NY).

Rocchi P. (2016). What information to measure? How to measure it?, *Kybernetes*, 45(5), 718–731.

Rocchi P. and Gianfagna L. (2005). An Introduction to the problem of the existence of classical and quantum information, *AIP Conference Proceedings*, 810, 248–258.

Rocchi P. and Panella O. (2021). A purely probabilistic approach to quantum measurement and collapse, *Advanced Studies in Theoretical Physics*, 15(1), 19–35.

Rocchi P. and Panella O. (2022). Methodological requirements and some basic issues of quantum mechanics, *Physics Essays*, 35(1), 32–38.

Rueckner W. and Peidle J. (2013). Young's double-slit experiment with single photons and quantum eraser, *American Journal of Physics*, 81(12), 951–958.

Scharf G. (2014). *Finite Quantum Electrodynamics: The Causal Approach*, 3rd edition (Dover Publications, Chicago).

Schaff J.F, Langen T., and Schmiedmayer J. (2014). Interferometry with atoms, *Rivista del Nuovo Cimento*, 37(10), 509–589.

Scheider W. (1986). Do the double slit experiment the way it was originally done, *The Physics Teacher*, 24, 217–219.

Scully M.O. and Drühl K. (1982). Quantum eraser: A proposed photon correlation experiment concerning observation and 'delayed choice' in quantum mechanics, *Physics Review A*, 25, 2208–2213.

Squires E. (1994). *The Mystery of the Quantum World* (Taylor and Francis Group, London).

Thomson G.P. (1928). Experiments on the diffraction of cathode rays, *Proceedings of the Royal Society A*, 117, 600–609.

Tonomura A., Endo J., Matsuda T., and Kawasaki T. (1989). Demonstration of single-electron buildup of an interference pattern, *American Journal of Physics*, 57, 117–120.

Veit J. and Solarek D.J. (1975). Interference fringes using a Fresnel biprism and a laser, *The Physics Teacher*, 13(7), 413–420.

von Neumann J. (1955). *Mathematical Foundations of Quantum Mechanics* (Princeton University Press, Princeton, NJ).

Weinert F. (2009). Davisson – Germer Experiment, In D. Greenberger, K. Hentschel, and F. Weinert (eds.), *Compendium of Quantum Physics*, pp. 150–152 (Springer, Berlin; New York).

Wheaton B.R. (1978). Philipp Lenard and the photoelectric effect, 1889–1911, *Historical Studies in the Physical Sciences*, 9, 299–322.

Whiteside D.T. (1970). The mathematical principles underlying Newton's Principia Mathematica, *Journal for the History of Astronomy*, 1, 116–138.

Xiao Y., Wiseman H.M., Xu J.S., Kedem Y., Li C.F., and Guo G.C. (2019). Observing momentum disturbance in double-slit "which-way" measurements, *Science Advances*, 5(6), eaav9547.

Zeh H.D. (2013). *The Physical Basis of The Direction of Time* (Springer, Berlin; New York).

Zeilinger A. (1999). Experiment and the foundations of quantum physics, *More Things in Heaven and Earth*, pp. 482–498 (Springer, Berlin; New York).

Zeilinger A. (2005). The message of the quantum, *Nature*, 438, 743.

Zeilinger A., Gaehler R., Shull C.G., Treimer W., and Mampe W. (1988). Single- and double-slit diffraction of neutrons, *Reviews of Modern Physics*, 60, 1067–1073.

Zhou Y., Peng T., Chen H., Liu J., and Shih Y. (2018). Towards non-degenerate quantum lithography, *Applied Sciences*, 8(8), 2076–3417.

Zurek W.H. (2003). Decoherence, einselection, and the quantum origins of the classical, *Reviews of Modern Physics*, 7, 715–775.

Chapter 18

Postface

Since my early studies on probability, I was inclined to believe that researchers' quarrels stemmed substantially from narrow intellectual horizons.

18.1.1 My first book was an attempt to outline a comprehensive view and chiefly emphasized the importance of the event as a theoretical starting point. For the most part, it presented the structural model that the work experience in software engineering inspired me in those years. That book was little more than a draft and Mazliak wrote: "[It] often retains the character of an essay on a work (or a thought) in progress. It will be interesting following possible future developments" (*MathSciNet*, AMS, 2006).

18.1.2 So, my second book began to fill the gulf. The theorems of large numbers and a single number prove that the frequentist and subjectivist probabilities descend from the number of the events under consideration. The work "does not side with either school but adopts a pluralistic approach, claiming that these two interpretations are in fact compatible, and that one can choose between them based on the applications at hand." (M. Bona, *Choice*, May, 2015). Holik added: "The ideas presented in the book open the door to interesting research inquiries in the field, especially those related to the foundations of quantum probability theory, quantum information theory and quantum statistical techniques." (*MathSciNet*, AMS, 2016).

Table 18.1. Attributes of probability deriving from events.

Events	Probabilities
Physical/Mental	Ontic/Epistemic
Deterministic/Indeterministic	Integer/Decimal
Disjoint/Combined	Summed/Multiplied
Testable/Untestable	Frequentist/Subjective
Influenced by material factors	Objective conditional
Influenced by cognitive factors	Subjective conditional

18.1.3 The reactions of the colleagues encouraged me to undertake this book, which gathers the reflections on probability developed over some 30 years.

The first part of this book aims to show how the masters constantly lean toward believing that the model of P is simple or can be simplified. Even when they become conscious of the complex texture of indeterministic logic, they presume that one or at most two models are enough to describe P. They forgo a comprehensive research plan and voluntarily quit the complete construction capable of explaining the thorough 'Geometry of Chance' announced by Pascal.

The second part of this book puts forward a conceptual framework that attempts to deduce the principal properties of probability from the notion of event. It does not demolish the theoretical achievements which have been discovered but strives to place them within an integrated framework.

This work adopts a ground-breaking approach. To the best of my knowledge, no probabilist has decided to infer the properties of $P(\mathbf{E})$ from $\mathbf{E}$ so far. Ten opposite features of probability, deriving from the concept of event, demonstrate the efficiency of the present method of study (Table 18.1). The main traits of frequency and subjective probabilities have been deduced from the number of experiments (Table 18.2).

The third part of the book explores the properties of the outcomes in classical and quantum environments. It is in accord with Popper who claimed the complete theory of probability is necessary to attack

Table 18.2. Main features of the frequentist and subjectivist probabilities.

Hypothesis	Theorems	Control of $P(A_n)$	Meaning of $P(A_n)$	Perspectives	Statistics
$n \to \infty$	TLN	Testable	Objective	Ontological	*Classical*
$1 \leq n < z$	TSN; UBT	Untestable	Subjective	Epistemic	*Bayesian*

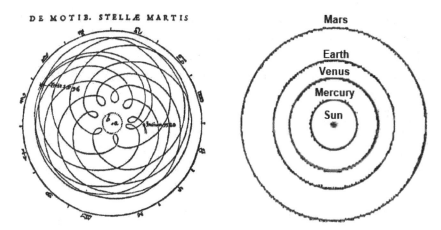

Figure 18.1. The orbit of Mars in the Ptolemaic system (left) and Copernican system (right).

the quantum foundations. This part also shows how the present theory can be validated or refuted by means of practical experiments.

Concluding, this book proposes a new approach and an innovative framework, in the effort to achieve a comprehensive and coherent understanding of probability. I personally believe that this attempt is not insignificant from the cultural stance (despite evident imperfections) because it points to a new direction to follow in mathematical research.

18.1.4 *An episode* – This book calls to my mind the following episode in the history of science.

Based on naïve visual observations, Ptolemy set up the cosmological model by placing the Earth at the center of the Universe.

Astronomers adhered to the *geocentric model* 'en masse' and excogitated complicated explanations to justify the strange movements of planets which appear to orbit in one direction and then turn back in the sky. The *heliocentric model*, formulated by Copernicus in *De Revolutionibus Orbium Coelestium (On the Revolutions of the Celestial Spheres)* (1543), solved these inexplicable issues with great precision (Figure 18.1), yet it met with the cold reception of coeval astronomers. Galilei sustained the thesis of the Polish scientist with vigor, Brahe inherited the Copernican ideas, and Kepler made the heliocentric model more understandable. The scientific community listened to the various proponents of Copernicus's scheme but did not change the idea.

By the end of the 17th century, Newton formulated the laws of mechanics and the universal law of gravitation, at that point, astronomers could but accept the correct planetary model. Almost 150 years had passed since the publication of Copernicus's book.

This book is the result of a lonely inquiry exactly like Copernicus's work.

I hope it will not wait for 150 years to be considered :-))))

Appendices

Appendix A – Textual Analysis: A Summary of Masters' Criticism

A.1 Richard von Mises (1957) devotes the initial part of the third chapter to the critical analysis of the opposing definitions of probability. He emphasizes the limit of the Laplacian scheme, which cannot be applied everywhere due to the equally likely constraint. He specifically holds that the subjective model may be influenced by psychological or physiological factors and puts down this model as follows: "The peculiar approach of subjectivists lies in the fact that they consider 'I presume that these cases are equally probable' to be equivalent to 'These cases are equally probable' since for them probability is only a subjective notion".

A.2 Leonard Jimmie Savage writes three pages in Chapter 4 of (1954) with disapproving annotations against the 'objective' interpretations of probability which do not have universal meaning. Savage summarizes his view: "... objectivistic views typically attach probability only to very special events. Thus, on no ordinary objectivistic view would it be meaningful... Secondly, objectivistic views are .... charged with circularity. They are generally predicated on the existence in nature of processes that may ... be represented by ... an infinite sequence of independent events."

A.3 Hans Reichenbach (1949) illustrates the historical evolution of the probability concept in the first chapter. Chapter 9 discusses the various meanings of probability; in particular, he examines the probability of a single event, which should have no place in science. Moreover, he charges: if the concept of probability only represents a subjective expectation, then probability does not have any connection to the real world.

A.4 John Venn (1888) criticizes the interpretation of probability as a personal belief, especially in relation to the thought of De Morgan in Chapter 6. He states: "the difficulty of obtaining any measure of the amount of our belief" and adds: "we experience hope or fear in so many very instances, that $\cdots$ whilst we profess to consider the whole quantity of our belief we will in reality consider only a portion of it." Venn concludes that actual human belief is one of the most elusive and variable factors so that we can scarcely ever get a sufficiently clear hold of it to measure it. Chapter 10 tackles another disputed argument, specifically it questions whether the events calculated by probability calculus are to be attributed to chance, on the one hand, or alternatively to causation or design, on the other hand.

A.5 John Maynard Keynes makes a historical retrospect of probability calculus in Chapter 7 of his treatise (1921). He illustrates the frequency theory recalling the work of Leslie Ellis and also looking into the Venn work. Chapter 8 emphasizes the limits of the frequentist model that clearly excludes a great number of judgments which are generally believed to deal with probability. Keynes also stresses the practical use of statistical frequencies, since "an event may possess more than one frequency, and that we must decide which of these to prefer on extraneous grounds." Later Keynes emphasizes the differences between Venn's construction and the generalized theory which he means to put forward.

A.6 Bruno de Finetti illustrates his theory in two volumes (1970) peppered with unfavorable and even sarcastic judgments about the opponent theories, especially the frequentist. In the first chapter he places the notions typical of the subjective and objective schools of probability side by side in order to highlight the profound differences extant between them. For instance, he notes that for 'objectivists', two events are independent if the occurrence of one does not affect the probability of the other; instead, for a subjectivist, two events are independent if the knowledge of one does not modify the probability assessment of the other event. Chapter 6 introduces three main interpretations of distribution functions, then the author begins a long discussion against countable additivity. Chapter 12 deals with estimations and testing that have distinct characteristics from the perspectives of the classical and the Bayesian statistics. De Finetti does not miss the opportunity to emphasize the Bayesian techniques and criticize the alternative methods here and there.

A.7 Frank Plumpton Ramsey begins the seventh chapter of the essay (1931) with censorious comments on the works of von Mises and Keynes. He pinpoints that the second author recognizes the subjectivity of probability, but in substance Keynes does not assign any value to subjectivism. Moreover, Keynes believes there is an objective relationship between knowledge and probability, as knowledge is disembodied and not personal. Ramsey has less severe judgments against the frequentist theory since he admits its adoption in science: "I am willing for the present to concede to the frequency theory that probability as used in modern science is really the same as frequency."

Appendix B – Semiotics in Brief

B.1 Premise

Writers, orators, and philosophers have investigated human communication for a long time. The earliest studies belonging to the Greek–Roman period dealt with specific themes such as writing, fine arts, grammatical rules, oratory abilities, and so forth.

In the essay 'Περὶ Ἑρμηνέας (On interpretation)', Aristotle (384–322 B.C.) looked at the relationships between language and logic. Specifically, he dealt with propositions that include the terms '*every*', '*no*', and '*some*', and explored the compatible or incompatible relations among them.

In the Middle Ages scholars meant to prescribe the rules necessary to make writing correct, beautiful, persuasive, true, and validly connected. They implemented the speech arts called *rhetoric*, *grammar*, and *logic*.

Besides the works which essentially pursued specific targets and abilities, a minority group had more general aims, in the sense that the authors sought the universal components of language and their intrinsic properties. Sextus Empiricus (160–210 B.C.) informed us that the Stoics began to delineate the basic elements of communication (Yngve, 1981):

> "Three things are linked together: what is conveyed by the linguistic sign, the linguistic sign itself and the object or event (···). Two of these are corporeal i.e. the sound and the object or event, and one is incorporeal i.e. the matter of the discourse conveyed by the linguistic sign."

In more recent centuries, Francis Bacon (1561–1626), John Locke (1632–1704), and other scholars methodically examined the anatomy of signs. They pioneered modern inquiries and, in a sense, envisioned *semiotics*, that is the discipline presently defined as the study of all kinds of signs — human and animal, stored and transmitted, verbal and postural, artistic and scientific, etc. — and of any process involving signs, especially the production of meaning.

Most authors ascribe the birth of semiotics to the Swiss Ferdinand De Saussure (1857–1913) and the American thinker Charles Sanders

Peirce (1839–1914). Working independently of each other, they inaugurated two distinct schools of thought.

From the 1960s onward, semioticians have attempted to harmonize the positions of Saussure, Peirce, and other authors, and have contributed to making semiotics a unified field. Various distinguished experts actively helped mature this domain of knowledge as a consistent discipline such as the American Charles William Morris (1901–1979), the French Roland Barthes (1915–1980), and the Italian Umberto Eco (1932–2016). In summary, it can be said that thinkers have been paying attention to the content of the semiotic subject since the dawn of time, but they organized the discipline of semiotics no earlier than 50 years ago.

B.2 Ferdinand De Saussure

The initial principles of semiotics date back to the lessons held forth by Saussure in Geneva in the years 1908–1909 and 1910–1911. The lecture notes, rearranged by the students Charles Bailly and Albert Séchehaye, have been published posthumously in 1916 under the title 'Cours de Linguistique Générale'.

Saussure establishes that the linguistic sign is equipped with two elements: '*le signifiant (signifier)*' and '*le signifié (signified or meaning)*'. For example, the noun 'tree' is the signifier and the 'idea of tree' is the signified, both placed in the human mind. These are the 'two faces' of the sign, which is basically an 'entité psychique (psychic entity)'. Saussure interprets language as a structure "où tout se tient (where everything stands bound)" at the level of the brain since signifiers and signifieds are mental entities.

He even stresses the idea that language is a dynamic system of interconnected units. For this reason, commentators associate the Saussurian thought with *structuralism*, a dominant approach by the mid of the 20th century.

The Swiss writer introduces the dimensions of linguistic analysis named *paradigmatic* and *syntagmatic*. The former indicates which words are likely to belong to the same category. *Lexical units* are defined, for instance, nouns and verbs. Syntagmatic relations enable

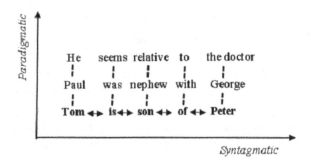

Figure B.1. Syntagmatic and paradigmatic directions of language development.

one to build up a formal proposition within *syntactical rules*. It may be said that the syntagmatic paradigm follows the direction of the discourse logic. Several scholars see the two paradigms as the axes of the language developments (Figure B.1).

One of the most important concepts of Saussure's lesson is the *principle of arbitrariness* which justifies the infinite variety of signs invented by people to communicate. This principle holds there is no privileged link between a signifier and its signified. No specific noun is more suited to a meaning than another sign. The semantic nexus turns out to be entirely discretionary and the principle of arbitrariness allows people to invent countless signs and related meanings.

This freedom hides the risk of creating so many words that communication between people becomes impossible to handle. Unlimited semantic freedom would prevent individuals from exchanging information, instead appropriate restrictions help individuals manage signs more easily. *Motivations*, which narrow down the variety of words and restrict the semantic extravagance, make the use of language more agile. For example, *grammatical motivations* establish the rules for getting the plural from the singular, for declining verbs, preparing augmentatives, diminutives, pejoratives, and other forms without coming up with new words.

People have invented broad assortments of motivations which apply *rational* and *irrational motives*; follow *criteria of convenience, social traditions*, tend to *help the memory*, etc. The study of motivations turns out to be somewhat complicated; for example, *etymology* is a large field of study devoted to the history of the word forms and, by extension, the origin and evolution of their semantic meaning over time.

Following the Saussurian lessons, one discovers a complex landscape where language is a historical and anthropological creation that the society of speakers sets up at a given moment in history. Individuals create their language as *an agreement*, by means of non-written conventions. In a way, Saussure anticipates Ludwig Wittgenstein (1889–1951) who claims that meaning is synonymous with use and denies the possibility of a "private language".

Various factors influence human life and, in turn, natural languages:

- *Synchronic linguistics* investigates languages as they are today.
- *Diachronic linguistics* conducts the evolutionary analysis of a language as a result of its changes over time.

The first discipline includes most of the works developed in the past centuries, for instance, grammar is a typical synchronic study. The second discipline follows the development and evolution of a language through history, either from the perspective of the present looking back to earlier stages, or from some earlier stage looking toward the present time.

B.3 Charles Sanders Peirce

Peirce and Saussure were both concerned about the basic tenets of semiotics, but their different personalities argued about diverse themes and produced diverse results. The first author was primarily a linguist, while the second was a multitalented philosopher, logician, mathematician, and scientist. Peirce investigated fundamental topics of mathematics and logic; he devoted himself to some aspects of epistemology and saw semiotics as the "formal doctrine of signs" closely related to logic.

Peirce offers a view of signs more complete; he holds that the sign "stands to somebody for something in some respect or capacity," so the signifier is not exclusively confined within the human mind

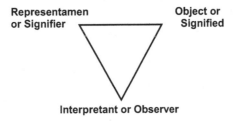

Figure B.2. The semantic triangle.

but pertains also to the real world; it can have a concrete base. In summary, the sign is a triangle (Figure B.2) endowed with:

- The *representamen* = The form, material or even immaterial, which the sign takes on (signifier).
- The *object* = The element to which the sign refers (signified).
- The *interpretant* = The physical observer or even the mental process, which relates to the other two elements.

The semantic triangle offers a reference in computer science that processes exclusively material signifiers.

Peirce believes that semioticians should not investigate the signs in isolation but as parts of the *sign system* that, on the one hand, is connected to communication and, on the other hand, to the construction of reality. The American author puts forward elaborate taxonomies of signs which in a way spell out the arbitrariness principle of Saussure. He establishes the following principal types of signs:

- *Symbol*: The signifier does not resemble the signified but is fundamentally arbitrary or purely conventional to the extent that its meaning must be learned. For example, numbers, most words of the language, traffic lights, and national flags are symbols.
- *Icon*: The signifier is perceived as resembling or imitating the signified (by looking at it, feeling it, hearing it, tasting it, or smelling it). It is similar in possessing some of its qualities. For example, a portrait, a cartoon, a scale-model, an onomatopoeic word, and mime gestures are icons.

- *Index*: The signifier is not arbitrary but is directly connected in some way – physically or causally – to the signified. This link can be observed or inferred. For example, smoke is the index of a fire; and knocks on a door indicate that someone has arrived.

Charles M. Morris, an American philosopher and semiotician, establishes the following main branches of linguistics, deriving from Peirce's criteria:

- *Semantics*: Is the study of the meanings of linguistic expressions. Besides the abstract and concrete nature of the represented objects, semantics explores the intention of the author, its truth value, its ambiguity, etc.
- *Syntax* or *syntactics*: Deals with the arrangement of words and phrases to create well-formed sentences in a language. It analyzes the set of rules, principles, and processes that govern the structure of sentences in a given language.
- *Pragmatics*: Has to do with the environment-dependent features of a specific language. It looks into the values of an expression which vary from context to context. For example, a single word has different meanings in different social ambits.

B.4 The State of the Art

This concise summary probably gives a false impression to the reader who underestimates the troubled road followed by semiotics in the 20th century.

The 'Course' of Saussure was well written but was mainly confined to the linguistic context. Peirce offered a broader view although his thought was not so easy to grasp due to the dispersive style of his writings. Some of Peirce's original ideas were not recognized until after his death, so a number of significant works are acknowledged posthumously. Moreover, Peirce often changed his views, and the texts must be consulted in the right order.

Several authors, coming from different countries, enriched the semiotic literature, but it was not easy for reviewers to synthesize the various viewpoints and systematize the different opinions. Major and minor writers have adopted personal terminologies and insisted on subtle conceptual differences to the point that commentators often had to mention the exact English, French, or German terms used by each author. The vast majority of writers has humanist extraction and is somewhat alien to mathematical methods; a certain cultural divide can be felt between semioticians and technology doers. The very concept of 'meaning' has very heterogeneous definitions such as:

- An intrinsic property,
- A unique unanalyzable relation of other things,
- The other words annexed to a word in the dictionary,
- The connotation of a word,
- An essence,
- An activity projected into an object,
- A volition,
- The place of anything in a system.

Even the name of the new discipline raised controversies: '*semiotics*', referring to the Peircean school, was put forward in opposition to '*semiology*' descending from the Saussurean tradition.

To end on a positive note: semiotics, which risked becoming a tower of Babel, gradually converged toward shared tenets during the second half of the 20th century, and currently continues to mature and consolidate. Semiotics has become a fertile domain, it investigates areas that turn out to be highly topical in the present society. Numerous fields of studies have sprouted with time such as the following:

- Biosemiotics,
- Semiotic anthropology,
- Zoosemiotics,
- Semiotics of new media,
- Ethnosemiotics,
- Marketing or commercial semiotics,
- Semiotics of arts (music, photography, cinema etc.),
- Social semiotics, and
- Urban semiotics.

Suggested Readings

Chandler D. (2017). *Semiotics: The Basics* (Routledge, London). Available as '*Semiotics for Beginners*' at http://visual-memory.co.uk/daniel/Documents/S4B/?LMCL=VsxorM.

Nöth W. (1995). *Handbook of Semiotics* (Indiana University Press, Bloomington, IN).

Deely J. (2004). *Basics of Semiotics* (St. Augustines Press, South Bend, IN).

Rastier F. (1999). *De la Signification au Sens: Pour une Sémiotique sans Ontologie*, Available at http://www.revue-texto.net/Inedits/Rastier/Rastier_Semiotique-ontologie.html.

Rocchi P. (2012). *Logic of Analog and Digital Machines*, 2nd revised edition (Nova Science Publishers, Hauppauge, NY). Partially available at http://www.edscuola.it/archivio/software/bit/course/book.html

Index